COMPUTERS AND THE ENVIRONMENT:
UNDERSTANDING AND MANAGING THEIR IMPACTS

ECO-EFFICIENCY IN INDUSTRY AND SCIENCE

VOLUME 14

The titles published in this series are listed at the end of this volume.

Computers and the Environment: Understanding and Managing their Impacts

Edited by

Ruediger Kuehr
United Nations University,
Zero Emissions Forum,
European Focal Point,
Hamburg, Germany

and

Eric Williams
United Nations University,
Environmental and Sustainable Development Programme,
Tokyo, Japan

KLUWER ACADEMIC PUBLISHERS
DORDRECHT / BOSTON / LONDON

A C.I.P. Catalogue record for this book is available from the Library of Congress.

ISBN 1-4020-1679-4

Published by Kluwer Academic Publishers,
P.O. Box 17, 3300 AA Dordrecht, The Netherlands.

Sold and distributed in North, Central and South America
by Kluwer Academic Publishers,
101 Philip Drive, Norwell, MA 02061, U.S.A.

In all other countries, sold and distributed
by Kluwer Academic Publishers,
P.O. Box 322, 3300 AH Dordrecht, The Netherlands.

Printed on acid-free paper

Cover:
"superfuture media; www.superfuture.com/supermedia"

Printed in the Netherlands.

TABLE OF CONTENTS

PREFACE

This book was conceived as a result of an autumn 1999 workshop of the United Nations University/Global Environment Information Centre (UNU/GEIC) and Professor Monte Cassim's group at Ritsumeikan University in Kyoto, Japan. We were introduced to the activities of a Japanese non-governmental organization that donates used computers to elderly and handicapped people to help them reap some of the benefits of the information technology revolution. One of us (Williams) has a background in industrial ecology and life cycle assessment, and had been researching the environmental implications of producing IT equipment. Another (Kuehr) came from an environmental policy background, and had been working on Zero Emissions, a UNU initiative working to mitigate environmental burdens by maximizing the utilization of materials. The activities of this NGO impressed upon us that managing the environmental impacts of computers is a much broader issue than we first thought. Through discussions with the workshop participants, we came to the conclusion that a multi-disciplinary "systems analysis" of the impacts and management of the computer life cycle, including use, re-use, and recycling, would be a project worth tackling.

Although it took longer to get here than we thought, this volume is the first major result of this effort—and it has been a bumpy road. We found that many of the issues to address had not yet been properly analyzed, so there was no existing pool of expertise to draw upon. Many of the contributors (including ourselves) ended up wrestling with the task of assembling a scattering of disparate crumbs of data to try to put together a useful analysis. Though there are clearly gaps remaining and much work is left for the future, we hope that the readers will find the result useful.

A number of issues that we originally wanted to cover are not included in the volume. In particular, we wanted to incorporate the implications of computers on the environment in industrializing nations. But in this area there were not even "data crumbs" to work with; addressing the issue requires creation of new primary data, a task beyond available resources. This is another challenge for future work.

We have struggled with the related issues of the audience for this volume and how technical it should be. In principle, the issues of computers and the environment touch everyone's lives, and thus should be interesting and accessible to anyone concerned about the environment. On the other hand, we also wanted to present an analysis that is relevant and convincing to those directly involved in analyzing and managing the environmental impacts of computers. This necessarily implies the use of more specialized ideas and tools. We ended up with a compromise: much of the material in the volume should be interesting for the general reader, but some is quite technical. When the treatment becomes abstruse, we have included non-technical summaries.

A preface is also a place for authors to thank those that have contributed to a work, and we would like to take the opportunity to do so. First and foremost, our gratitude goes to Dr. German T. Velasquez, who is not only responsible for finding much of the funding to undertake the work, but also brought in the idea to tackle attitudes and behavior of consumers and to consider the symbiotic relationship between hardware and software markets. His contribution goes far beyond the individual chapters he co-authored. We express special thanks to Professor Motoyuki Suzuki, former vice-rector at the United Nations University, who encouraged us and created a work environment where our efforts could proceed well. His bottom-line perspective of the big picture (e.g., he asked us how to define a "green PC") helped keep us on course. Thanks also to Randy Helten and his team (Greg Helten for proofreading and Susan Juby for layout) for their meticulous efforts. Many thanks also to our publishers at Kluwer Academic Publishers, Henny Hoogervorst, her assistants Amber Tanghe-Neely, Esther Verdries, and Gloria Verhey, for providing us a forum for insights and the views expressed here. The Japan Foundation-Center for Global Partnership helped to support the work financially through a grant for the "Digital Economy and the Environment" project from 2001–2003. There are many others who have contributed; some, but not all, are mentioned in individual chapters.

Our hope with this volume is to make a contribution to understanding and managing the environmental impacts of computers. In order to improve on our future work, we are interested in hearing your opinions on this effort. Readers wishing to contact us or to see more about the environmental implications of the information technology revolution are invited to visit the project Web site <www.it-environment.org>.

Eric Williams and Ruediger Kuehr
Tokyo (Japan) and Stockholm (Sweden)
July 2003

Chapter 1

COMPUTERS AND THE ENVIRONMENT—AN INTRODUCTION TO UNDERSTANDING AND MANAGING THEIR IMPACTS

Ruediger Kuehr,[a] German T. Velasquez,[b] Eric Williams[c]

[a] *United Nations University, Zero Emissions Forum, Germany*
[b] *United Nations University, Global Environment Information Centre, Japan*
[c] *United Nations University, Japan*

1. THE RISE OF PERSONAL COMPUTERS AND INFORMATION TECHNOLOGY

In just two decades, personal computers (PCs) have become ubiquitous in the homes and offices of the industrialized world. Manufacturing, sales, management, medicine, etc.,—all depend on computers now to function efficiently. E-mail has become indispensable in our day-to-day communications with family members, friends, and colleagues. It is now hard to imagine life (in rich countries) without computers in one form or other. Despite the rise of a variety of new devices to deliver information services, such as the Internet-capable cell phone, there is no obvious substitute on the horizon for the key features of a PC: large display, input keyboard, and personal information processing and storage capability.

PCs are a key component in the infrastructure of information technology (IT). Much as the internal combustion engine and electricity shaped the first half of the twentieth century, IT is a technological revolution driving change in the economies and societies around the world. One key aspect is the direct contribution of IT-producing sectors to national economies. In 1999, world production of equipment[1] was U.S.$1.09 trillion (OECD 2002), 2.5 percent of gross world product. In the United States, the Department of Commerce estimates that in 1998, 8 percent of GDP and 35 percent of

[1] The IT hardware sector here includes computers, office equipment, radio and telecommunications equipment, consumer electronics and components.

Computers and the Environment: Understanding and Managing Their Impacts
Edited by Ruediger Kuehr and Eric Williams, pages 1–15.

growth in income were due to the sectors producing IT hardware, software, and services (Henry 1999). Japan and the European Union are also centers for production, and in the industrializing world hardware industries in Korea, China, Singapore, Malaysia and Taiwan accounted for 23 percent of world production. India (mainly in the Bangalore area) is increasingly becoming a center for the software industry. Production of IT-related goods outside of the United States, Japan, the European Union, and Southeast Asia is limited. In addition, application of IT contributes indirectly to economic growth through making business processes more efficient.

IT also affects society to the extent that pundits often describe the changes as a new "digital society." One issue gaining particular attention is the digital divide—the growing inequality within and between nations due to differences in ability to take advantage of IT.

2. COMPUTERS AND THE ENVIRONMENT

In contrast with the economic and social aspects, the environmental implications of IT have not yet been subjected to a similar level of consideration and debate. As with any major technological revolution, the effects of IT for environmental sustainability are significant and wide ranging. Examples include increases in environmental efficiency in products and services (e.g., automated control of heating and lighting of buildings), shifts in transport and consumption patterns associated with use of e-commerce and telecommuting, and extra consumption stimulated by increased incomes/lower consumer prices (Romm et al. 1999; United Nations University 2003).

There are very direct environmental implications associated with PCs due to the impacts of production, use, and disposal of the equipment itself. This is an issue of consequence, especially given the huge number of computers in homes and offices. In April 2002 the billionth personal computer was shipped (Tech-Edge 2002), global annual production was some 113 million machines in 2000. PC penetration rates in the industrialized world are high, and it is plausible that the number of PCs will likely increase to one or more machines per capita in these countries. Meanwhile, growth in PC use is also rapid in much of the industrializing world.

Continued, rapid technological progress in the IT industry has contributed to short lifespans that are well below the functional limits of computers. A short lifespan exacerbates environmental impacts, requiring production of more new machines and increasing the numbers heading for landfills or recycling centers. Discarded products from the IT sector show the highest growth rate among municipal and industrial wastes. The total volume of IT products disposed of in Germany reached some 110,000 tonnes (30,000 private, 80,000 commercial) in 2000 (bvse 2000). In the same year the mountain of PC waste in Japan reached 73,000 tonnes, most of them commercial desktop PCs, followed by privately owned desktop PCs. In the United States, some 20.6 million PCs became obsolete in 1998 (EHC 1998). Only 11 percent of those were recycled and about 75 percent were put in storage (EPA 2000). According to a

Carnegie Mellon University study, by 2005 about 680 million PCs will have been sold in the United States since 1981. This study estimates that about 325 million will be older than five years in 2005 and thus defined as obsolete (Matthews et al. 1997).

The internal structure of the PC is complex, making recycling difficult, yet many of the machines are still either usable or contain usable components, creating a real challenge for end-of-life processing. Also, complexity implies that the production of PCs and their components is energy- and material intensive. Results of a recent study suggest that 1.3 kilograms of fossil fuels and chemicals are required to produce a two-gram memory chip (Williams et al. 2002). For a desktop PC the figure likely runs into the hundreds of kilograms of fossil fuels. Energy use to operate computers is also relevant. Recent analysis of the IT infrastructure in the United States indicates that PCs accounted for 1 percent of national electricity consumption in 1999 (Kawamoto et al. 2000).

Companies and governments have been trying to address the issues discussed above. Computer and component manufacturers continue to work towards increased energy and materials efficiency; part of the drive in this direction is no doubt due to the relevance of these issues for production costs. For the use phase, voluntary initiatives such as the Energy Star program of the U.S. Environmental Protection Agency and the Department of Energy has played an important role in improving the energy efficiency of computers and a host of other products (Weber et al. 2002). Many nations have already legislated mandatory requirements (or are in the process doing so) to institutionalize the collection and recycling of used equipment ("take-back" programs) The European Union recently enacted two directives related to waste electronics (WEEE and ROHS)—one requires recycling and the other restricts the use of particular substances, such as lead solder, in electronic equipment. At the national level, Belgium, Denmark, Italy, Japan, the Netherlands, Norway, Portugal, Sweden, Switzerland, and Taiwan have legislated mandatory guidelines for the recovery of electronic equipment. In the United States, a multi-stakeholder group known as the National Electronics Product Stewardship Initiative (NEPSI) is working towards the establishment of a financing system to deal with used electronic products.

3. PERSPECTIVE AND TARGET OF THIS VOLUME

Given the broad range of activities described above, what do we intend to contribute to the debate on these issues with this edited volume? At the general level, the main objective is to synthesize a multi-issue "systems picture" of the environmental impacts and management of the life cycle of PCs. The description should cover a broad range of key elements, including production and recycling technology, environmental analysis, corporate strategies, the re-use of PCs, consumer attitudes, and government policy. The need for such a synthesis is evident—information on individual issues is often scattered and none has yet to synthesize the different aspects.

In addition to this general objective we also have a more specific agenda. In the process of studying the various initiatives to deal with end-of-life computers, we could not help but notice that the issue of recycling dominates the discourse. While collection systems and recycling technologies are clearly needed, the common wisdom of waste management is that reducing and re-using waste streams are often very effective and economical approaches compared to recycling. This is often referred to as the "three Rs" (3Rs) approach, a strategy that prioritizes reduce, re-use, and recycle options in exactly that order.

To briefly explain the theoretical underpinnings of the 3R strategy, note from Figure 1 that there are number of options for dealing with waste streams. Recycled materials (e.g., aluminum cans) replace virgin materials needed for producing goods near the beginning of the overall supply chain—thus creating a loop in the material flow from the "grave" back to the "cradle." The re-use of goods can be considered as a flow back to the distribution and sales stage, and optimizing usage reduces demand for new and re-used goods in the first place. In general, the further back one goes in the production chain, the more processing (with associated environmental impacts) is required. Thus "shorter" loops tend to be more effective from environmental and, often, cost perspectives.

A significant share of the environmental impacts incurred in the computer life cycle is in the production phase. For a desktop computer used at home, for example, the energy needed to produce the machine is four times more than that needed to power it during the use phase. The energy consumed to produce a refrigerator is only about one-eighth the electricity used to run it.[2] Thus one would expect that using the same computer longer is especially effective at reducing total impacts. Also, more of the environmental burden associated with producing a PC comes from the making of high-tech parts than from producing the bulk materials like steel and aluminum. Thus, the environmental "payback" from materials recycling is comparatively less than for appliances like refrigerators; the fact that PC recycling generally does not turn a profit reflects this reality.

Given this context, a major objective of this publication is to explore how end-of-life options, in addition to recycling, can contribute to reducing the life cycle impacts of personal computers. When compiling this book, a special effort was made to sharpen the discussion beyond the generic terms reduce, re-use, and recycle. Computers have their own characteristics as products and thus deserve special consideration as such.

For one thing, PCs are also quite expensive, being second after only automobiles in price among widely-used consumer durables. Also, rapid technological change implies that they become obsolete very quickly. These factors suggest that two forms of end-of-life management may be very important: reselling and upgrading. Reselling is essentially about matching users with appropriate machines at optimal cost; when a computer no longer satisfies the needs of its user, she or he can sell or donate it to

[2] For most appliances the energy needed for production is much smaller than the electricity needed to operate the appliance.

another user who has lower demands for performance. Upgrading involves replacing specific components that underwent the most "technological progress" so as to maintain a similar level of performance compared to newer machines. Microprocessors, memory, and hard drives are typical examples of components suited for upgrade. There are obvious limits to the extent to which either option can be implemented—demand for used machines equipped with Pentium 386 processors is low, to say the least. Upgrading very old machines to a modern level would involve replacement of nearly everything but the keyboard and power supply, making extensive upgrades an unattractive option for users. The real question is this: How significant are the opportunities for reselling and upgrading, and are they being exploited sufficiently?

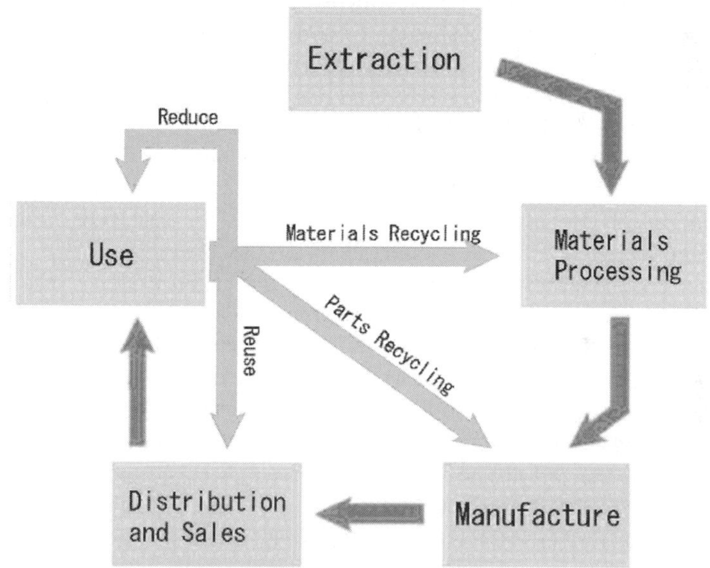

Figure 1: Back-flows from the use phase: reduce, re-use, recycle.

It is also important to consider how end-of-life management systems for computers relate to industrializing countries. Using information technology is important for economic growth, but the high cost of establishing an IT infrastructure is a major obstacle in less-wealthy nations. Thus, an international supply of used computers offers a means to reduce their costs in establishing IT infrastructure. There is the danger, however, that import of used computers by industrializing nations becomes an environmental burden instead of an economic boon. The expense involved in proper recycling of PCs implies that the export of used equipment can be an attractive "waste-treatment" option for the industrialized world. Such exports pose significant environmental risk if recycling processes are carried out without proper caution in the receiving countries. The groundbreaking report, *Exporting Harm: The High-Tech Trashing of Asia,* lays out a litany of environmental disasters that followed as a result of the establishment of a computer recovery and recycling industry in the town of

Guiyu, China (Puckett and Smith 2002). Although the impacts of such industries in industrializing countries deserve serious attention, we were not able to provide a detailed analysis of that topic here due to the extensive resources that would be needed to gather primary information and data.

4. STRUCTURE AND SUMMARY OF CHAPTERS

This book aims to shed light on the following specific issues:

- The environmental impacts incurred when producing PCs
- Electricity consumption in the use phase
- Environmental impacts of disposing of computers in landfills
- How green design of PCs can reduce environmental impacts
- Industrial perspectives and activities of leading computer companies
- Economic, managerial, and technological aspects of recycling
- The role of consumers in influencing the supply of green PCs and implementing environmentally friendly end-of-life options
- Evaluation of the environmental effectiveness of reselling, upgrading, and recycling computers
- How used-PC markets can extend the product lifespan
- How governmental (both national and regional) and non-governmental policy initiatives deal with the environmental management of PCs

In Chapter 2, Matthews and Matthews provide an expanded context for discussions in this book. The authors begin with a review of the rapid growth in the production and adoption of PCs and also point out that recent years have seen the rise of new "small" IT products such as the cell phone. In addition to the direct impacts of the production, use, and disposal of these products the ways in which IT is used by society and business also significantly affect the environment. The adoption of new IT will probably lead to efficiency improvements but, at the same time, also contribute to economic growth and structural changes that could negate the benefits of efficiency gains. The authors discuss how the disposition (i.e., storage, landfilling, and recycling) of end-of-life equipment plays an important role in the direct environmental impacts of used PCs. In particular, the tendency of users to store or stockpile equipment for several years reduces their re-usability and creates additional challenges for recycling. The chapter concludes with a survey of corporate-level approaches towards realizing more effective recovery and recycling of equipment, such as materials selection, proper labeling of parts, and take-back programs.

In Chapter 3, Williams takes on two tasks. One is to survey the environmental impacts associated with the production (and disposal in landfills) of computers. The second is to undertake an analysis of the energy, chemicals, and water used in the chain of manufacturing processes involved in making a computer. Williams identifies three main areas of environmental concern: energy use, possible long-term health effects of exposure and emissions from chemical-intensive production stages, and impacts from the leakage in landfill sites of hazardous materials contained in computers. There has been much debate on possible increases in cancer rates and birth defects relating to workers in semiconductor fabrication, but few studies have been done to clarify matters. Lead (in CRT monitors and circuit boards) and brominated flame retardants (in computer cases and circuit boards) are potentially harmful to humans and ecosystems, but here, as well, the potential impacts are still unclear. For lead, it seems that no one has yet done a risk assessment, despite an upcoming ban on lead in circuit boards in products sold in the European Union. The potential health risks of brominated flame retardants have only recently come to light (around 1999), and the medical community is still working to understand the nature and scope of the problem.

The result of the analysis of materials use revealed that to make one desktop computer at least 240 kilograms of fossil fuels, 22 kilograms of chemicals, and 1,500 kilograms of water are used. Computer production is materials-intensive; the total fossil fuels used to manufacture one computer, for example, amount to nine times the weight of the actual computer, compared to only one to two times for an automobile or refrigerator. Also, the production of a computer takes the lion's share of total energy use in the product life cycle—80 percent for manufacture versus 20 percent for electricity to run the computer—in contrast to other household appliances such as the refrigerator, where 88 percent of total energy use goes to running the device.

Future courses for the practical implementation of the European Union's new directive on waste electrical and electronic equipment (WEEE) are the focus of Chapter 4 by Hieronymi and Schneider, who compare two future scenarios for electronics recycling centers in the year 2018. One is high-tech, with sophisticated systems to automatically detect and separate different parts for re-use and recycling. In this scenario, there is a high degree of integration in the design and management processes of major firms, and a high proportion of recovered parts goes back into new products. In the second scenario a recycling plant uses processes that have remained more or less unaltered for fifty years. A varied mix of equipment is fed into uniform processes that shred and crudely separate electronics into several different mixed streams. While some materials are recoverable, other valuable materials cannot be separated without great cost, and problems arise over how to dispose of the unusable remainder. The authors suggest that the evolution towards one or the other of these scenarios depends on how the WEEE Directive is executed in practice. Implementation by a monopolist organization with no concern but to comply with the minimum requirements for "recycling" at fixed fees leads towards the second scenario. To move towards the first scenario, they argue for a model where a (pluralistic) organization works closely with industry and all actors are financially motivated to continuously improve the system to increase profitability. Negotiations and political processes in the near-term will determine which organizational model is adopted for future implementation.

In Chapter 5, Höhn and Brinkley present an examination of IBM's environmental management of various aspects of products, which includes a wide spectrum of activities and processes ranging from product design, supply chain management, and manufacturing to sales operations and customer relations management. In their view, the best way to manage the environmental aspects of a product is through the integration of environmental requirements in the product design, and from there into all other relevant business processes. This requires an environmental management system that includes elements to consider product characteristics. A main lesson learnt is that harmonization and standardization will be essential parts of global approaches for managing the environmental aspects of products.

The environmental management of another global player in the PC industry, Fujitsu Siemens Computers, is reviewed by Podratzky in Chapter 6. This chapter is largely dedicated to the company's efforts to reduce environmental impacts by focusing on the individual stages of their product's life cycle. Approaches taken include their corporate environmental policy, management systems, marketing of green products through eco-labels, and take-back, resale, and recycling systems. The first "Eco-PC," produced in 1993, and a currently-sold product using a lead-free soldered system-board are also described in detail as practical examples of fulfilling these requirements.

The implications of the rise of the PC on electricity use are examined by Cole in Chapter 7. The operation of office and telecommunication equipment is thought to account for around 3 percent of total electricity use in the United States, a substantial fraction of which is due to PCs. The chapter analyzes the structure and trends of electricity use of PCs. The "full-on" power consumption of computers shows conflicting trends. Increases in the power of microprocessors and monitor size tend to raise energy demand; but on the other hand, liquid crystal display (LCD) screens and laptop computers use less electricity than their counterparts and are growing in popularity. In general, technical advances in the industry, including miniaturization, tend to increase energy efficiency. The power management capabilities of computers can be as important as, or even more important than, "full-on" consumption. PCs are often left on for large blocks of time when not in use, even overnight. Automatic standby and sleep modes can thus significantly reduce energy use. These features are increasingly being incorporated and improved in computers and monitors, though obstacles remain, such as failure to use these functions and incompatibility with software and peripherals. Shutting off the monitor (and the computer itself) is effective for saving energy, but many users are reluctant to turn off equipment, because they have heard that frequent switching will cause harm. Apparently, the negative effects only appear if frequently switched on and off far beyond the effective lifetime of the device (e.g., frequent switching reduces the lifespan of a monitor only after 20 years of use).

In Chapter 8, Saied and Velasquez examine how consumer behavior relates to the "greening" of PCs. Through consumers' decisions on which computer to purchase, whether to use power management functions, how long to keep the PC, and finally, what to do with it at the end of its life, they significantly influence the environmental performance of computers. The use of eco-labels—systems that certify certain computer models as being environmentally preferable—is a key tool for including the environment in consumer's purchasing decisions. The variety of existing eco-labels is reviewed, from the widely-adopted Energy Star to the as-yet unused E.U. Flower. Surveys indicate that private and business users, by-and-large, do not yet consider environmental issues when choosing a PC. Saied and Velasquez emphasize that a key issue here is a lack of information, not only about eco-labels but also on how consumers should use and dispose of computers in an environmentally friendly way. How consumers would actually act on the information is not yet clear, but knowledge is at least necessary, if not sufficient, for action to be taken. They suggest that a priority be set on dissemination, through the popular media as well as with "environmentally friendly user" guides explaining the eco-labels a given machine is certified under, how to use power management, upgrade the machine, and where to send it for recycling.

An environmental assessment comparing different end-of-life options for computers is the topic of Chapter 9 by Williams and Sasaki. The premise is to explore the extent to which the reselling and upgrading of computers are environmentally effective options compared to recycling. Research reveals that reselling or upgrading one in ten computers reduces total energy use by 8.6 percent and 5.2 percent, respectively, by reducing demand for new machines. In contrast, recycling the materials from one in ten computers only saves 0.43 percent by replacing demand for virgin materials. The reason for this dramatic difference is that much of the energy in the life cycle of a computer is invested in its complex components (e.g., microchips) rather than bulk physical substance (e.g., aluminum frame and plastic case); thus the environmental returns from recycling are less than with many other goods. This finding provides evidence that the extension of lifespan should receive more serious attention in the policy arena addressing end-of-life computers.

As extending the PC's lifespan is apparently beneficial for the environment, in Chapter 10, Williams and Kuehr analyze the status, trends, and obstacles for used-PC markets. The markets for used PCs in the United States and Japan are significant and growing, with small- and medium-size firms, donation agencies, Internet resellers and brokers, and computer manufacturers playing various roles (adequate information on European or other markets was not readily available). This sector faces a variety of "non-economic" obstacles that should be addressed. One is the issue of the transfer of software licenses for pre-installed software to the purchaser of a used PC. Although a software license is, in principle, transferable to the person buying a used PC, it often is not simply because the mechanics of the transfer are needlessly awkward. The problem can be fixed through relatively simple changes in how licenses are packaged and some degree of user awareness. Also, many consumers are probably not aware of how to utilize secondary markets to either find a home for their older computers or a source of inexpensive computing power to fit their needs.

Klatt analyzes the recycling of PCs in Chapter 11. The basic flow of the recycling process is transporting, sorting, and dismantling machines, separating of parts into categories, and then processing these parts into different streams for appropriate handling. Steel and aluminum in the cases of computers are relatively easy to recover. Copper is common and relatively valuable, but difficult to recover as it is contained in coated wires and dispersed in circuit boards. Circuit boards contain small amounts of precious metals such as gold, silver, and platinum; recovery requires refining to separate these materials. A cathode ray tube (CRT) monitor contains several kilograms of high-lead content glass, but a major difficulty in recycling is the efficient separation of the different types of glass used. Re-usable parts can also be recovered, but this is difficult to implement in practice with existing systems. The economics of computer recycling make it only marginally profitable. Some recyclers can cover costs if clients pay for shipping to the recyclers, suggesting that net costs to recycle a computer, including transport, is in the tens of dollars. The economics of PC recycling vary substantially depending on factors such as local logistics, current market conditions, recycling technology used, and the material and parts composition of the equipment. Low labor costs in industrializing nations can lead to net profits for recyclers there, but the environmental risks of low-tech processing are extreme (as evidenced by the case of the town of Guiyu, China).

In Chapter 12, Sarkis examines the operations of a computer equipment resource recovery facility, providing insight into how one actually works. One aspect that becomes clear is that a "demanufacturing" facility shares many of the management characteristics of one engaged in manufacturing—supplier and customer relations, need for planning, control tools, and flow lines, for example. This chapter looks at the issues surrounding the "inputs" at the Compaq recovery facility, including the requirement of purchasing materials from suppliers who, at the same time, are also customers. This includes some of the operational characteristics of the facility, including flows and operational process steps, especially those that focus on source reduction and environmental management. Included in this discussion is the role of the facility's workers and the work environment itself. The chapter describes appropriate training aimed at limiting risk and improving process efficiency, which are important since many of the workers are relatively unskilled and on (temporary) contracts. The chapter concludes with a discussion on the outputs of the recovery facility, including some of the issues related to the development of partnerships with potential customers and relationships with suppliers of waste management and recovery services.

In the book's final chapter, Kuehr focuses on policies to address the PC's life cycle. He compares the policies of Japan, the United States, Germany, and the European Union in order to identify their political priorities, under the hypothesis that governmental approaches under current policy sets are ineffective in preventively mitigating the environmental impacts of IT products. This chapter also examines the effectiveness of certain industrial and non-governmental approaches. It becomes apparent that present policies are most concerned with recycling, but do not really do much to address other key issues, such as the extension of product life, reduction of harmful effects during manufacture, energy consumption during usage, and the reduction of material inputs. Kuehr sees a possible way out of this dilemma through the development of taxation instruments that reduce the advantage of new over old machines, and continued dematerialization by fostering services that offer equipment leasing and renting (Robèrt et al. 2002). The chapter also shows that discussions on these topics are still dominated by an economic perspective and mainly curative countermeasures (ie. reactive vs. proactive), and that the general situation is still a far-from-promising with respect to achieving sustainable development.

5. THE FUTURE AGENDA TO REALIZE GREEN PRODUCTION, USE, AND DISPOSAL OF COMPUTERS

Collectively, the chapters in this volume present the state of research and knowledge on some, but by no means all, of the issues related to the environment and PCs. Overall, what do they suggest in general terms about how societies should approach the environmental management of computers in the future? To elaborate, we first lay out a framework showing how different societal actors (governments, companies, and civil society) are involved in managing the environmental impacts of computers.

Government (national, regional, and local) roles include the following:

- Environmental regulation of manufacturing processes (e.g., emissions from semiconductor fabrication facilities)
- Regulating environmental characteristics of products (e.g., banning the use of lead)
- Legislating mandatory take-back and recycling systems
- Introducing voluntary programs such as eco-labels and the Toxics Reporting Initiative[3]
- Setting taxation rules (e.g., depreciation rates for valuing new and used computers)
- Funding education and public awareness activities
- Funding and implementing research and analysis
- Setting ground rules on issues such as transfer or export of hazardous waste, means of transferring software copyrights, etc.

Companies (manufacturers, leasers, resellers, and recyclers) have tasks such as these:

- Designing environmentally-friendly computers
- Controlling the environmental management of manufacturing processes
- Reporting firm-level environmental data to the public and/or government
- Communicating other environmental information to customers (product documentation, Web site, advertisements)
- Implementing take-back and recycling activities
- Acting as mediators in secondary markets

The roles of some actors in civil society include the following:

- Non-governmental organizations (NGOs)—acting as agents facilitating the flow of donated PCs, and expressing the environmental concerns of citizens

[3] The Toxics Reporting Initiative (TRI) is an initiative of the United States Environmental Protection Agency that surveys companies on emissions of a wide variety of toxic chemicals.

- Academia—engaging in research, analysis, and education
- Consumers—providing a market for environmentally-friendly PCs; utilizing power management features of computers; using computers longer before purchasing new machines; and influencing the end-of-life flow of PCs to used-product markets, recycling centers, or landfills.

Also, international organizations are involved in analysis (e.g., this publication) and in setting international standards, such as the ISO 14000 series of environmental management standards. It is through the collective actions of all these groups that the environmental impacts of computers will be reduced.

Given this structure of actors and roles, what directions do the chapters in this publication suggest for an appropriate set of future actions?

5.1 Research and Analysis

Because the computer is still a relative newcomer to the world stage, there remain many unanswered questions regarding its environmental impacts and management. While a lack of complete knowledge is no justification for inaction, it is also difficult to plan an appropriate path in an information vacuum. This implies the need for an agenda of research and analysis. Some research topics of particular relevance are listed below:

1. Effects of chemical use on IT workers and local ecosystems in both industrializing and already-industrialized countries
2. Further work to understand environmental impacts at the equipment level (per computer, for example) to guide policy priorities
3. Estimations of potential macro-level impacts of waste IT equipment (e.g., effects of soldering lead in landfills on local water quality)
4. Surveys of the needs, wants, practices, and attitudes of computer users
5. Extension of product lifespans: clarification of status, potential, and obstacles for expansion of used markets and the practice of upgrading

Academia, government organizations, and NGOs are key players in undertaking the needed analysis, and cooperation with industry is important for addressing many issues.

5.2 Information and Knowledge Creation and Dissemination

Data is the fuel that drives the engine of analysis. The above agenda cannot be pursued without the creation and public availability of new information. Some points, such as understanding the needs and practices of users, are easily resolved, simply requiring some group with sufficient resources to carry out the needed large-scale surveys. Others, such as the health effects of brominated flame retardants, challenge the frontiers of medical knowledge. The answers will hopefully come in time, presuming scientists continue to pursue the key questions. Another class of data-related problems is the public availability of information on industrial processes. This is a thorny issue, especially in high-tech sectors where firms are, understandably, reluctant to divulge information. In the European Union, Japan, and elsewhere, consumers are being asked to bear the economic burden of recycling computers; yet there is not a single publicly-available set of data that describes the environmental characteristics (such as energy use) of recycling processes. Here, the actions of industrial organizations and government agencies can do much to address the situation. These are both well positioned to collect and aggregate corporate-level data for public use without divulging the individual sources (Williams 2000).

The dissemination of knowledge is also important, in particular with respect to the civil sector. Chapters 8 and 10 raise the issue that users are by-and-large unaware of what actions they can take to reduce the environmental impacts of computers. Some of these options, such as prompt sale to secondary markets, are also beneficial for users, so they are likely receptive to such information. Some aspects, such as the meaning of the different eco-labels and use of power management functions can easily be incorporated into documentation included with machines. Other issues, such as how to take advantage of upgrades or used-computer markets, may be more appropriately disseminated via the popular media.

5.3 Setting Standards for the "Green PC"

What, indeed, is a "green PC"? There are many possible paths for future computer technology to take, and it is desirable to choose the most environmentally-friendly sub-paths whenever feasible. Visions of and setting standards for the green PC can play a useful role in taking the right paths in practice. Eco-labels and legislation of product characteristics (e.g., bans on the use of certain substances) are two important ways of setting standards for green PCs. While some eco-labels have been effective at stimulating improvements, two issues stand out as important for their future development. One is the evaluation of the environmental benefits gained by using the label. The process by which eco-label criteria are decided is often unclear and is rarely supported with analysis that justifies why certain choices were made and information about the expected benefits of those choices. We thus suggest a process for developing eco-labels that includes both input from a broader range of actors and analysis of the choices made.[4] Also, a plethora of different regional eco-label standards has sprung up—a situation that is awkward for both manufacturers and consumers. The establishment of a standard international eco-label would be very helpful in resolving these issues.

The ban on certain substances being used in PCs (such as lead in solder) being introduced by the European Union has aroused controversy and criticism, particularly from the United States. Critics claim there is a lack of scientific evidence justifying the ban, and apparently there are no studies "waiting in the wings" to counter this argument. The substance bans may indeed be a very good idea but, as with eco-labels,

[4] The process could resemble an environmental impact assessment (Glasson et al. 1994).

they should be developed in concert with analyses that evaluate what environmental risk they really reduce.

In addition to what a green PC contains and how it performs, the standards for a green PC should also address how it is made. For example, should processes to make chips be free of ozone-depleting substances? Currently, the main direction of ensuring environmental friendliness in production is to ensure that one's own facilities and suppliers are compliant with the ISO 14000 environmental management standards. This probably has a positive effect, but overlooks the fact that the ISO management standards only create a *system* for improving environmental performance; they do not touch on actual environmental performance itself. Production issues could be more directly addressed using an eco-label that deals specifically with PC production issues.

5.4 Optimizing the PC as a Service, Not Just as a Machine

Notwithstanding the points made in the previous section, it is crucial to think beyond the machine itself and consider how to maximize the environmental performance of the *services* delivered by PCs. A recurring theme throughout this book is the short lifetime of a computer, which greatly exacerbates its environmental impacts. The simplest and most effective way to reduce environmental burden may be to ensure that users need fewer new PCs in the first place. A variety of governmental policies as well as actions by corporations and civil society could work to achieve this. Two examples mentioned in this book include ensuring the smooth transfer of software licenses to secondary users and enhancing tax breaks to encourage users to donate their used PCs. There are many other possibilities. The main task ahead is to consider the "toolbox" of possible policy approaches, keeping in mind the goal of maximizing the *service* gained from PCs. This approach will go far in mitigating the environmental impacts of personal computers.

REFERENCES

bvse. 2000. *Elektro- und Elektronikschrottverwertung* (in German). Bonn: bvse.

Environmental Health Centre (EHC). 1998. Electronic Product Recovery and Recycling Conference. Summary report.

Environmental Protection Agency (EPA). 2000. WasteWise update: Electronics reuse and recycling. <www.epa.gov/wastewise> (20 March 2003).

Glasson J., R. Therivel and A. Chadwick. 1994. *Introduction to environmental impact assessment.* London: UCL Press.

Henry, D., Sandra Cooke, Patricia Buckley, Jess Dumagan, Gurmukh Gill, Dennis Pastore and Susan LaPorte. 1999. *The emerging digital economy II.* Washington, D.C.: U.S. Department of Commerce.

Kawamoto, Kaoru, Jonathan G. Koomey, Bruce Nordman, Richard E. Brown, Mary Ann Piette, Michael Ting and Alan K. Meier. 2000. Electricity used by office equipment and network equipment in the U.S.: Detailed report and appendices. Published in proceedings of the 2000 ACEEE Summer Study on Energy Efficiency in Buildings, August 2000, at Asilomar, CA. Berkeley: August Ernest Orlando Lawrence Berkeley National Laboratory.

Matthews, H. Scott, Chris T. Hendrickson, Francis C. McMichael and Deanna J. Hart. 1997. *Disposition and end-of-life options for personal computers*. Pittsburgh: Carnegie Mellon University (CMU). <www.ce.cmu.edu/GreenDesign/comprec> (26 November 2002).

Organisation for Economic Cooperation and Development (OECD). 2002. *OECD information technology outlook: ICTs and the information economy*. Paris: OECD.

———. 2002b. *OECD in figures*. Paris: OECD.

Puckett, Jim and Ted Smith, eds. 2002. Exporting harm: The high-tech trashing of Asia. Report released by the Basel Action Network and Silicon Valley Toxics Coalition, 25 February 2002.

Robèrt, K.-H., B. Schmidt-Bleek, J. Alosi de Larderel, G. Basile, J. L. Jansen, R. Kuehr, P. Price Thomas, M. Suzuki, P. Hawken and M. Wackernagel 2002. Strategic sustainable development—Selection, design and synergies of applied tools. *Journal of Cleaner Production* 10:192–214.

Romm, J., A. Rosenfeld and S. Herrman. 1999. *The Internet economy and global warming*. Annandale: Center for Energy and Climate Solutions. <www.cool-companies.org/energy/> (8 September 2002).

Tech-Edge. 2002. 1 billion served, PCs go over the top, and into the dump. <http://homepage.mac.com/techedgeezine/1billion_served.html> (10 November 2002).

United Nations University. 2003. Information technology and environmental initiative. <www.it-environment.org> (February 2003).

Weber, C. A., R. B. Brown, A. Mahajan and J. Koomey. 2002. *Savings estimates for the Energy Star voluntary labeling program: 2001 status report*. Berkeley: Ernest Orlando Lawrence Berkeley National Laboratory.

Williams, E. 2000. Global production chains and sustainability: The case of high-purity silicon and its applications in information technology and renewable energy. Monograph. Tokyo: United Nations University. <www.it-environment.org/publications/> (March 2003).

Williams, E., R. Ayres and M. Heller. 2002. The 1.7 kg microchip: Energy and chemical use in the production of semiconductors. *Environmental Science & Technology* 36 (24) (15 December): 5504–5510.

Chapter 2

INFORMATION TECHNOLOGY PRODUCTS AND THE ENVIRONMENT

H. Scott Matthews and Deanna H. Matthews
Carnegie Mellon University, USA

1. INTRODUCTION

Information and communication technologies (ICTs) have become critical components of global infrastructure over the past few decades, and computers are now fundamental to most business processes. The global adoption of the Internet has only accelerated the transition from physical to digital infrastructure. More generally, the use of information and communication technologies has resulted in far-reaching changes in production processes and product characteristics.

While the global economy has generally been expanding for the past decade, largely due to growth in high technology, there has been relatively little attention focused on the potential environmental impacts of the widespread deployment of ICTs around the world. The negative environmental impacts of ICTs arise from a globally-polluting supply chain, producing the electricity needed to power the world's installed stock of computers, and a rapid obsolescence pattern which leads to discarded equipment. However, ICTs also have the potential to generate significant environmental benefits through changes in product design and production processes, and more generally, through the more efficient use of resources. Understanding the combination of these positive and negative factors is critical in assessing the "green-ness" of personal computers.

As a result, concern is growing globally about how to manage the volumes of ICT equipment. These products historically have relatively short life spans, and end-of-life management is an important tool towards minimizing the volume of products being sent to landfills. Other options, such as reusing, remanufacturing, and recycling, exist and need to be promoted.

This chapter seeks to introduce the reader to the scale of the issues involved in the proliferation of information technology equipment, the history of disposition

Computers and the Environment: Understanding and Managing Their Impacts
Edited by Ruediger Kuehr and Eric Williams, pages 17–39.

assessment models and the difficulties in following these issues, and to describe some of the alternatives to disposal. The scope of the chapter extends beyond computers to include wireless telephones, personal digital assistants (PDAs), and other popular electronic products. The concept of life cycle assessment is introduced to motivate a holistic approach to analyzing environmental issues. Finally, scenarios are constructed to estimate the current stock of electronic equipment around the world and flows of equipment with various end-of-life alternatives.

2. THE IMPORTANCE OF INFORMATION AND COMMUNICATION TECHNOLOGIES

The OECD[1] report, *Measuring the ICT Sector*, provides a broad overview of the importance of the ICT sector across many member countries (OECD 2000). Some of the report's key findings are summarized below.

2.1 Economic Scale of ICT Sectors

Of the roughly 13 million persons employed by the sector worldwide, 35 percent are from European Union (E.U.) countries, 39 percent from North America, and 16 percent from Japan. In addition, on a country-by-country basis, employment in ICTs as a percent of total business sector employment is between 2 to 6 percent, with an OECD average of 3.6 percent. This suggests that the countries most directly benefiting from increases in ICT growth are developed nations. The implications, however, for other countries further down the ICT supply chain may be considerable, but are less clear.

Another way to look at the economic impact of ICTs is to estimate value added, which is defined as the difference between the value of a final product and the sum of intermediate inputs. Value added from ICTs across all countries considered was estimated to be U.S.$1.2 trillion in 1997. About half of that value was created by the United States (U.S.). Further, the G7 countries contributed 87 percent of total value added. In total, value added represents nearly 45 percent of the total value of production in the ICT sector. For individual countries this value ranged from 28 to 51 percent. Overall, nearly 10 percent of value added in the ICT sector is spent on research and development.

Schreyer (2000) estimated the average annual rate of growth of constant price expenditures on information and communication technology equipment for seven countries for the periods 1985–1990 and 1990–1996. Table 1 summarizes these results.

These average annual growth rates doubled output in some sectors from 1992 to 1997 in the U.S., while overall U.S. economic output increased by only 35 percent for the same period. Thus, growth in ICT sectors is having a disproportionate impact on growth rates.

---

[1] Organization for Economic Cooperation and Development

Table 1: Average annual growth rates of expenditure on ICT equipment (percent).

Period	Canada	France	W. Germany	Italy	Japan	U.K.	U.S.
1985–1990	17.2	16.2	18.8	20.8	23.6	25.5	19.6
1990–1996	17.6	11.0	18.6	12.9	14.5	17.6	23.8

Source: Schreyer (2000).

2.2 Rise of the Personal Computer

When the IBM personal computer (PC) was put on the market on 12 August 1981, it introduced the world to the concept of having a computer on every desk. While it was primarily aimed at businesses, some consumers were also attracted to having a computer in their home. It was not the first personal computer developed, but it was the one that changed the way society accepted computers as a part of everyday life. Unlike computers introduced by other companies, it had an open architecture, meaning it was assembled from available parts and sold at many different outlets around the world, such as computer stores, mass retailers like Sears, etc. It ran on a 4.77 megahertz (MHz) Intel 8088 microprocessor and came with 16 kilobytes (KB) of memory, which could be expanded to 256 KB. You could buy models with either one or two 160 KB floppy disk drives and a monochrome monitor. The price tag started at $1,565 (about $3,100 in year 2001 U.S. dollars), which suggests that real prices over time for computers have remained relatively constant. Meanwhile, $3,000 in 2003 could buy a 2 GHz computer with 256 MB of memory, and 50 gigabytes of storage.

Sales of personal computers exploded. While only about 300,000 "desktop" computers were sold in the U.S. in 1980, 1.5 million were sold in 1981, and nearly 3 million in 1982. Since then, the U.S. and the world have generally seen an exponential growth of computers. A summary of worldwide personal computer sales from 1981 to 2002 is shown in Figure 1. While sales stalled in 2001 and have remained flat since, overall world demand is expected to grow between 8 and 10 percent in the short term. Regardless, over 130 million computers are being sold per year around the world. By the end of 2002, one billion PCs had been sold worldwide. Many of them, however, became obsolete years ago.

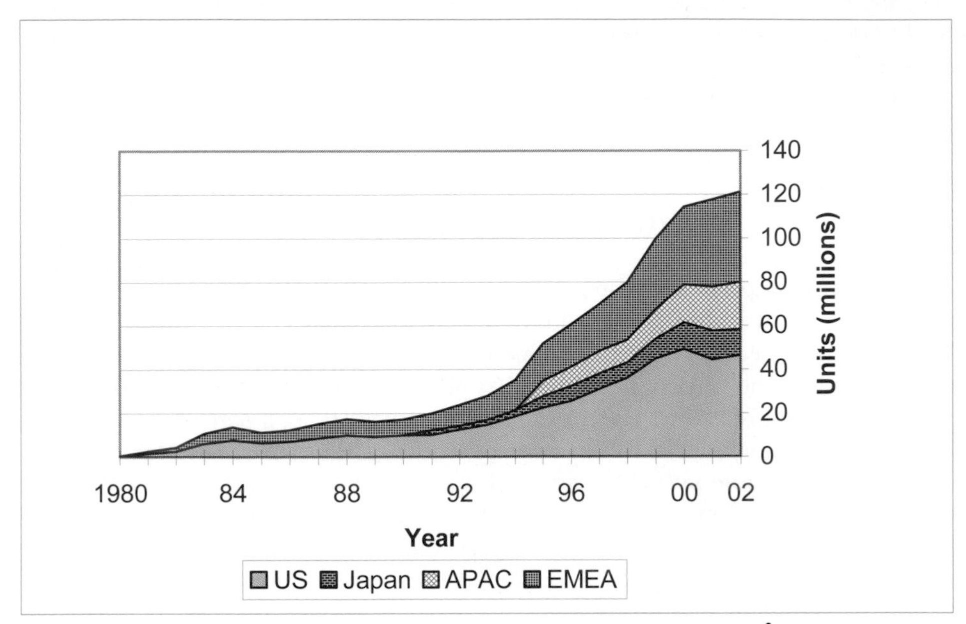

Figure 1: Annual U.S. and global computer sales, 1980–2002.[2]
Source: Dataquest (1996–2003).

Similarly, Table 2 summarizes recent estimates of global Internet usage per capita. As of 2002, the U.S. leads with almost 25 percent of the world's 665 million Internet users; however, it is only ranked seventh in Internet users per capita, behind the Scandinavian countries.

Around the world, about 10 percent of people had Internet access by the end of 2002. This estimate suggests that the vast majority of the world population does not have access to the Internet, and is most likely receiving little direct benefit from ICT growth. As there are no reliable data on global PC ownership, Internet access data is a fair proxy for ownership.[3] Note, though, that many of these people only have access via mobile phones.

2.3 Other ICT Equipment

In 1990, only about ten million people worldwide subscribed to wireless telephone services. Growth has been about 50 percent per year since 1996—with the one-billion-subscribers mark crossed in early 2002 (ITU 2001). Growth rates are especially high in developing countries like China. Handset sales for the past few years have been

---

[2] APAC refers to sales in the Asia-Pacific region. EMEA refers to Europe, Middle East, and Africa.

[3] For more on global PC ownership, see <http://cyberatlas.internet.com/big_picture/geographics/article/0,,5911_429391,00.html>.

increasing to keep pace with subscriber growth, but subscribers continue to rapidly replace their old handsets. In 1999, about 300 million wireless telephone handsets were sold worldwide, and another 400 million a year from 2000 to 2002, an indicator that subscribers are frequently upgrading their equipment to take advantage of the newest features—at a rate even faster than PC replacement.

These numbers are highly relevant, as there are now more people with wireless phones than fixed wired phones, with fixed wired lines also at just over one billion (ITU 2001). Wireless telephone users will far outnumber PC users, given that just under one billion total PCs will have been sold by the end of 2001 (and many of those were obsolete long ago).

Table 2: Top 15 countries in Internet usage.

	Internet users (thousands)	**Share %**
1. U.S.	160,700	24.13
2. Japan	64,800	9.73
3. China	54,500	6.71
4. Germany	30,350	8.18
5. U.K.	27,150	4.08
6. South Korea	26,900	4.04
7. Italy	20,850	3.13
8. Canada	17,830	2.68
9. France	16,650	2.50
10. India	16,580	2.49
11. Brazil	15,840	2.38
12. Russia	13,500	2.03
13. Australia	10,450	1.57
14. Spain	10,390	1.56
15. Taiwan	9,510	1.43
Top 15 Total	496,000	74.48
Worldwide Total	665,910	100

Source: Internet Industry Almanac (2002)

According to the Cellular Telecommunications Industry Association, as of 2002 there are currently about 140 million mobile telephone users in the U.S. and about 130,000 antennas for mobile telephone networks (CTIA 2001). The vast majority of the communication via these antennas (a.k.a. wireless towers) is analog or digital voice traffic. Emerging global standards for wireless communication, however, will allow devices of many kinds to share wireless bandwidth. As these wireless networks are deployed, there will be a need for much smaller "cells" (area covered by one set of transceiving equipment) to handle the increased bandwidth load. As a comparison, current cell sizes for mobile telephone networks are on the order of five square kilometers. Future "micro-cells" could be as small as 0.01 square kilometers. Moving towards this kind of system will necessitate massive infrastructure investments, and generally, replacement of existing equipment. In addition, improved technology and lower prices have sparked a soaring demand for wireless services.

A distinct possibility in the short term about growth in this sector revolves around consumer migration towards mobile and handheld computing platforms. This likely migration is seen partially by the movement towards portable computing (e.g., laptops) in the developed world. In 2000, while sales of desktop systems were generally flat, sales of laptops increased 20 percent. There has been a distinct movement in the past several years towards laptop and notebook computers, especially in the corporate market, as the performance penalties long associated with laptop systems, compared to desktop systems, have shrunk. Since 1995, laptop sales have increased from 30 percent of total personal computer sales in the U.S. and Japan to 50 percent in Japan and over 40 percent in the U.S. This cannibalization of desktop computers by laptops is expected to continue, and is forecast to affect other world markets.

At the same time, sales of handheld computing devices (e.g., portable digital assistants, or PDAs) tripled. It seems clear in the short term that computing power has hit a "sweet spot" in that consumers are seeing small marginal benefits from increased desktop computing power, but see large benefits from devices they are able to take with them. Growth in sales of these devices is high—2.5 million units were sold worldwide in 1997; 12 million were sold in 2002. PDAs allow users the base functionality of electronic life planners, to hold memos and contact information of acquaintances, etc. Applications exist, however, that allow file transfer, electronic mail access, Internet surfing, etc. Current and next generation mobile telephones are also being designed to replace traditional computing devices. These devices allow Internet access and can also act as PDAs.

In the developed world, a "convergence" phenomenon with smaller devices replacing larger devices is being seen. Thus, any scenario considering future ICT deployment patterns should realize the potential for widespread adoption of these smaller devices as a computing alternative in the rest of the world. While only some potential and current PC users can substitute handhelds, ignoring these markets—which are clearly part of the sector—could seriously underestimate economic and environmental implications.

In the long run, it is difficult to predict whether sales of small devices will be substitutes for larger equipment. The availability of mobile communications infrastructure, however, is a key requirement for this to happen. It is also possible that sales of smaller devices will simply represent additional demand for ICTs. An interesting historical analogy is the cannibalization of typewriter sales by personal computers. For example, IBM began selling typewriters in 1935, became a major player in that market, and introduced the PC in 1981. By then, word processing machines had already begun to replace typewriters. By 1990, word processing was a standard software package on computers. Nowadays, typewriters are scarce in offices—kept around only to facilitate quick printing of envelope labels and paper forms. The same cannibalization effect is happening to other historic office machines, such as cash registers and fax machines.

It is also important to note, however, that there are many areas of the world "under-served" by mobile communications. These areas could choose to deploy large or small wireless cells, depending on the estimated demand for bandwidth. This fact will have implications on the usage of ICT products. Since the world has reached the "cross-over" point for wireless versus wired telephone subscribers, and continued penetration of subscribers is expected worldwide, most future new telephone users, especially in developing countries, will receive wireless service.

While ownership and unit sales of ICTs are expected to continue their dramatic increase in the next decade, only a small fraction of the world's population are currently direct users of ICTs, although they may, of course, be affected indirectly in a number of ways. In fact, less than a third of the world's population has ever even made a telephone call. Recent evidence in developed countries suggests that computing devices are becoming more portable (e.g., the movement from desktop to laptop computers and portable digital assistants).

3. ENVIRONMENTAL IMPLICATIONS OF INFORMATION AND COMMUNICATION TECHNOLOGIES

In this section, some of the environmental implications of ICT production and use are reviewed. In the first sub-section, the environmental effects of production are reviewed, followed by two sub-sections, which look at energy use and waste disposal. It is important to bear in mind that these sections discuss the "gross" negative effects and do not seek to review the potential substitution effects, which are likely to result in less important "net" negative effects, or even positive effects. These depend upon substitution effects. For instance, if energy use for communication substitutes for energy use for transport, the net environmental implications may well be positive.

3.1 Direct and Indirect Environmental Effects of ICT Production

Life cycle assessment (LCA), shown in Figure 2, is a systematic approach that links a product's life cycle, from design to disposition, with the environmental impacts generated at each stage. A life cycle assessment consists of three complementary components—inventory, impact, and improvement—and an integrative procedure, known as scoping. Results in this section are divided into life cycle stages to ensure that all relevant stages of ICT equipment and services are represented.

Europe has put more resources into performing LCA, in large part due to the establishment of the Environment Directorate (DGX1) by the European Commission. European approaches to LCA were done in parallel to those developed in the U.S. by the Society for Environmental Toxicology and Chemistry (SETAC) (Fava 1994). The U.S. Environmental Protection Agency (EPA) accepted and built on the SETAC framework (EPA 2000; Vigon et al. 1993; Curran 1996).

The first step of an LCA consists of stating the goal and appropriately identifying the boundaries of the study. For example, in the LCA of a paper cup, the analyst might decide to consider only the ten highest cost items in the production process to save time and effort (Hocking 1991). The boundary definition is important, since it separates what will be considered in the analysis from what will be excluded. Thus, the boundary definition could lead to significant under-estimation of the discharge inventory. The other components are as listed by Vigon et al. (1993). The inventory analysis quantifies the energy, raw material requirements, and environmental discharges for the entire life cycle of the product or service. The impact analysis characterizes the effects on the environment and human health of the resource requirements and environmental loadings identified in the inventory stage. Finally, the improvement analysis systematically evaluates the needs and opportunities to reduce the impacts on the environment and human health associated with the life cycle of the product.

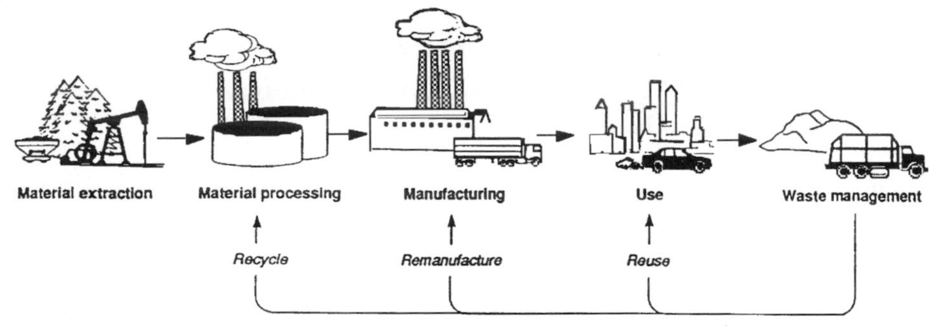

Figure 2: Life cycle assessment involves analysis of the entire supply chain.
Source: U.S. Office of Technology Assessment (1992).

On a per-unit basis, the study in Chapter 3 of this volume estimates that the final production of a desktop computer (including material production and manufacture) requires at least 240 kilograms (kg) of fossil fuels, 22 kg of chemicals, and 1,500 kg of water. Water use in silicon wafer production is decreasing but still significant. Assuming a chip size of one square inch (6.5 square centimeters), a survey of international SEMATECH[4] member companies shows that the average amount of water used in production on a per square inch of silicon basis dropped from 22 gallons (83 liters) to 17 gallons (64 liters) after study recommendations were implemented. The International Technology Roadmap for Semiconductors (ITRS) called for a reduction to 13 gallons (49 liters) per square inch of silicon by 2001 for a state-of-the-art semiconductor fabrication facility (SEMATECH 2000). Given the average production of a semiconductor factory of two million chips per month, this results in a water consumption of 20 million gallons (76 million liters) per month that must be managed and treated before being returned to the environment. The environmental implications will, of course, depend upon the availability of raw water supply and the characteristics of the receiving environment.

---

4 SEMATECH is a global consortium of semiconductor manufacturing companies.

3.2 Resource Use and Waste Generation from ICT Equipment

ICT use also has implications for resource inputs other than energy. Those of the most significance from an environmental perspective are the metal and materials sectors. While there is a continued migration towards plastic, rather than metal parts, the critical functions of this type of equipment are achieved by advances in uses of metals and materials. Metals still used in ICTs are aluminum, lead, cadmium, and nickel. Extraction of these metals can, of course, result in a disruption of natural landscapes.

One of the greatest challenges in managing the explosive growth of ICT equipment in the past decade has been (and will continue to be) the management of end-of-life assets. It is the deployment and replacement of equipment that should be a primary concern for environmental impacts. As technological expansion continues, more and more products will become obsolete. Existing studies on the fate of computers have suggested that while technological options exist to recycle obsolete electronic equipment, the majority of such equipment is stored in warehouses or in households. Regardless, the ultimate sink of such efforts—which are not 100 percent efficient—continues to be landfills. As the volume of ICT equipment grows, the useful lives of products decrease, and technological changes force upgrades, the volume of equipment going to landfills will increase. Some particular concerns are the expected transition to digital television, mobile telephones, and portable digital assistants. Digital television broadcast standard phase-ins will abruptly create hundreds of millions of old televisions in the U.S.

Some of the specific environmental concerns of ICT products are the heavy metals in computers and monitors and the varied toxic substances contained within. Flame retardants are another important issue, as they introduce hazardous substances into the supply chain. Moreover, as noted above, the manufacture of ICTs depends upon many other industries, including metal fabrication. Aluminum itself is neither toxic nor hazardous, but is made by processing ores from the ground. Other metals, though, are toxic to humans. For example, lead exposure is known to lead to developmental problems in children (i.e., reduced intelligence quotient [IQ] scores), and was phased out of gasoline in many parts of the world beginning in the 1970s. Some of the materials used (e.g., silicon) are either toxic themselves or rely upon highly toxic processes to fabricate them.

One reason the implications of landfilling ICT equipment are so large is due to the sheer volume of equipment. Since 1981, over 400 million personal computers have been sold in the U.S. (Matthews et al. 1997; National Safety Council 1999). As of 1998, there were six million tonnes of waste from electrical and electronic equipment (WEEE) produced per year in the E.U., accounting for 4 percent of municipal waste (CEC 2000). When electronic products are dumped in landfills, the materials have the potential to leach into the environment. Leaching occurs when a substance "leaks out" from its product into the environment. This happens to electronic products in landfills because of the natural "decomposition" process that occurs. For example, lead has historically been used in cathode ray tubes for computer monitors. A University of Florida study showed that leachate levels of lead from computer monitors are up to three times higher than leachate standards in the U.S. (Townsend et al. 1999). This is not surprising, as many studies fail to find evidence of large-scale recycling or remanufacturing (Matthews et al. 1997; National Safety Council 1999). For example, in 1997 and 1998, only about ten million units per year were managed through computer recycling companies in the U.S. Studies continue to show that owners of ICT equipment overestimate its value, and tend to store it, rather than reselling or recycling.

Indeed, a recent American study by Matthews et al. (1997) estimated that a much larger fraction of obsolete equipment is stored than recycled. The authors estimate that 140 million computers will be landfilled in the U.S. by 2005. It seems that computer owners do not comprehend the rapidly decreasing value of their equipment, and make sub-optimal decisions regarding its fate. A similar study, conducted by the Japanese Electronic Industry Development Association (JEIDA) in 1997, found that despite increasing productivity of recycling technologies, a relatively low volume of equipment was actually being recycled. If these trends continue, large quantities of electronic equipment will need to be managed in the future. Other studies have painted a more optimistic picture. For instance, a U.K. report estimated that one million metric tons per year of WEEE is created each year (0.25 percent of total waste), that 90 percent of such waste is large ICT equipment, and that roughly 50 percent of this waste is recycled (ICER 2000). Ferrous metals were estimated to represent 40 percent of this waste.

Another waste implication of ICT equipment is the increasing use of portable energy sources (e.g., batteries). While many products contain rechargeable batteries, most rely on disposable batteries. Often the rechargeable batteries are nickel-cadmium based; both metals have high toxicity and should be diverted from waste streams due to the highly efficient recycling rates of these substances. Another concern pertains to the miniaturization and discounting of ICT products. As products become smaller and less expensive, consumers will most likely feel more comfortable with simply disposing of these products.

3.3 Indirect Effects of ICTs on Environmental Issues

3.3.1 Rebound and Scale Effects

As noted by Berkhout and Hertin (2001), there are potential implications from ICT equipment as a result of structural economic changes. While ICTs have been seen to have impacts on reducing the resource intensity of economies, the general shift from manufacturing to service industries has not been without environmental costs (Rosenblum et al. 2000). Part of the reason for these environmental costs can be attributed to scale and rebound effects.

Scale effects are associated with increased investment in ICTs, because of their potential effects on productivity, and thus aggregate demand in the economy. Thus, environmental impacts due to the increase in the scale of economic activity can outstrip the positive impacts from any potential efficiency effects. The rebound effect, which is a more micro-level phenomenon, is another area of concern with regard to ICT products. This effect occurs when, for example, energy efficiency gains are more than offset by increases in use—thus the energy use "rebounds" above the original level. Due to generally falling prices for ICT products, there is concern that there will be adverse environmental implications from increased direct consumption, use, and disposal of the sector's outputs.

In short, scale and rebound effects are at the center of the problem when analyzing the prospective environmental implications of ICTs. While the extent of future environmental impacts from investment and use of ICTs is unknown, it is likely that some of these expenditures will inevitably lead to increased resource or energy use, waste, and emissions. It will remain important for society to closely monitor the environment to determine these cases and consider corrective action.

Another potential option is "extended producer/product responsibility," since this allows policymakers to address environmental problems "upstream," before the post-consumption phase. Currently, there are few structured options available to consumers of ICT products at end-of-life. This is significant, because, if left unchecked, this equipment could potentially end up in landfills. In Europe, the Waste Electric and Electronic Equipment (WEEE) Directive is currently being finalized and will most likely take effect in 2004. The goal of this directive is to divert large amounts of electronic product waste from the municipal waste stream. Manufacturers would be responsible for "taking back" used electric and electronic products (of which ICT equipment is only a subset of what also includes items like power tools and appliances). When realized, the WEEE directive will be the most comprehensive and wide-ranging initiative to manage ICT products in the world.

An important step in the U.S. is the recent Consumer Education Initiative started by the Electronics Industries Alliance (EIA). In February 2001, the EIA introduced a web site (www.eiae.org) that directs owners of ICT equipment to companies and organizations that accept old equipment, such as charities and remanufacturing companies. While the information and resources presented are significant, it remains unclear whether this "awareness campaign" will lead to increased amounts of re-use and recycling. Without clear links to economic incentives (such as vouchers for shipping and handling of equipment), there is little benefit to consumers to participate in such programs. Regardless, the EIA Initiative is a step in the right direction for managing such waste.

The most common fate for obsolete ICT equipment at present, however, appears to be ending up in landfills. This presents an environmental concern for several reasons. First, as mentioned above, the volume of equipment could be significant. Current estimates of obsolete personal computer systems in the U.S. alone suggest the volume to be on the order of 200 million cubic feet (5.7 million cubic meters), or roughly the volume of a football field stacked 1.5 kilometers high. Further, this number does not include other large ICT products like televisions. Televisions present a similar and interesting case because of the expected transition around the world from analog to digital television equipment. As this transition occurs, there will be a period of "forced obsolescence" for consumers, who will need to buy new equipment, and the market for used televisions will be small.

4. STOCK AND FLOW ESTIMATES OF COMPUTER LIFE CYCLE PATHWAYS

A 1991 Carnegie Mellon study predicted that 148 million PCs would find their way into U.S. landfills by the year 2005 (CMU 1991). The study relied on three fundamental assumptions: ten years of historical U.S. sales data, with a forecasted 5 percent future growth rate; a simple, one-stage model of end-of-life disposition; and some predictive assumptions on the future disposal rates of computers. An updated model created in 1997, more representative of conditions at the time, included more end-of-life options and estimated that 150 million computers would be recycled (and only 55 million landfilled) by 2005 (Matthews 1997). Finally, another estimation model created by the National Safety Council (NSC) surveyed electronics recycling firms and estimated significantly less computer recycling than in the 1997 model (National Safety Council 1999).

The remainder of this section suggests some scenarios with which to analyze the current stocks and flows of computer hardware equipment in the U.S. and worldwide. These scenarios are intended to motivate concern for end-of-life management issues, specifically, to minimize the volumes of equipment being sent to landfills or incinerators. Since actual data on the recycling of equipment does not exist, these scenarios represent only estimates of the magnitude of obsolete equipment and should not be construed as actual volumes.

4.1 A Stock and Flow Model

The arrows in Figure 3 define the pathways of computers in our updated, multi-stage model. A new computer is purchased, and eventually becomes obsolete. At that time, there are four options available to the owner of the computer. First, it could be re-used. This means that it is somehow used again after becoming obsolete to the purchaser—possibly a result of being resold or reassigned to another user without extensive modification. Second, the original owner could store the computer. In this case, it is serving no purpose except to occupy space. Third, the computer could be recycled. This is defined as meaning that the product is taken apart and individual materials or sub-assemblies are sold for scrap. Finally, the computer could be landfilled. In our model, the re-use and storage options are only intermediate stages in the life cycle of a computer (although anecdotal evidence suggests storage might be indefinite in some cases). Only recycling and landfilling are terminal points. Notice that our model allows a computer to be purchased, re-used, stored, and finally, recycled or landfilled.

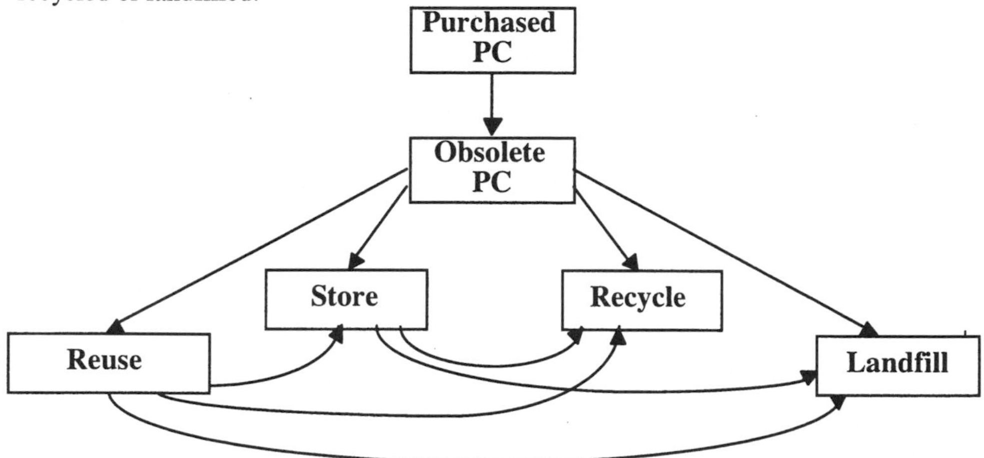

Figure 3: Flow diagram of computers in updated study.

Since the initial estimates were derived several important changes have occurred. The single most significant change is the creation of many computer "recycling" firms. However, the "recycling" done by such firms often falls into the categories of both re-use and recycling as defined in the original model. Before proceeding further, these two end-of-life activities are defined. Re-use implies that the computer is put back into service intact or with very minimal modification (such as adding memory). This could happen as a result of the computer being resold to a different user, being reassigned to a different user within a firm, or simply continuing to be used in another capacity by the purchaser. Recycling implies that the computer has in some way been disassembled and sold for raw materials or as separate electronic components.

Computer "recycling" firms both recycle and re-use products. Some firms actually take obsolete computers, make slight modifications, and place them in schools, non-profit organizations, and charities for extended lifetime use ("re-use"). However, some firms actually "recycle" electronic equipment, extracting value from the components or high-value materials like gold and other precious metals.

The emergence of these industries has diverted the flow of many computers from the municipal waste stream, however, hard numbers are not readily available which measure the aggregate numbers of computers being diverted into the re-use and recycling industries. In addition, consumers and businesses alike have shown a considerable unwillingness to throw away old electronic products. Instead, computers are often stockpiled in attics and storerooms until space is needed for another purpose. The existence of adequate storage space has also contributed to the diversion of computers from landfills, as owners have had time to put off the disposition decision long enough for the recycling and re-use markets to mature.

Other important changes are that the percentage of laptop computers sold has been rising, such that nearly 50 percent of new PC sales in the U.S. and Japan are laptops. This has the potential to reduce the landfill volume of computers but, of course, increases the difficulty of recycling, since laptops are smaller and components more concentrated. Finally, as seen in the National Safety Council's report, the incidence of recycling computer systems was less than previously estimated. The NSC report estimated that roughly three million PCs were recycled in 1997 (as opposed to the CMU estimate of seven million). Thus, estimates of recycling have been lowered to meet this surveyed data, and the number of units recycled per year more closely meet these industry estimates.

The factors shaping the disposition of computers today are important in building an updated model of their fate. Following the three fundamental points of the initial study, the sales data are updated to 2002 (from Dataquest sources and press releases—see Figure 1) and assume a more conservative growth rate of 6 percent per year, going forward. Finally, as mentioned previously, the actual number of computers being re-used, recycled, stored, and landfilled is not available, thus estimates of these activities are made by extrapolating from individual recycling firm data as well as making qualitative assessments of the level of activity.

The final update is in suggesting the percentages of computers that follow the various end-of-life options. It is assumed that computers generally become obsolete to the original purchaser in four years. This is one of the few assumptions carried over from the original model, and is supported inside and outside the industry. Market research firms like Dataquest and International Data Corporation (IDC) use a 3.5-year lifetime to predict product and processor upgrade paths in the PC industry. If not landfilled or recycled after becoming obsolete, re-use and storage allow for three more years in that capacity, implying the products are seven years old when leaving that step in the process. Re-used computers can be recycled or landfilled after seven years; or they can be stored for three more years before then being landfilled or recycled, for a total life of ten years.

The first level of assumptions requires us to determine the percentages of obsolete computers being re-used, recycled, stored, and landfilled. Since no firm data on this is available, two observations were used. First, information from articles and interviews with computer recycling firms shows few computers less than five years old are being processed. Thus, recycling and landfilling at the top level were assumed to be almost negligible. In addition, experience with corporate and individual users indicated roughly a one-to-one choice between re-use and stockpiling after four years. Thus, it is assumed that the fate of the remaining obsolete computers is split evenly between the two options. Second, as mentioned above, the NSC report gave estimates of the number of computers recycled per year, which are internally validated in this model.

Re-used machines are defined as seven-year-old equipment now of no use to the owner. They are either recycled, stockpiled, or landfilled. Again, the percent landfilled is assumed to be small. Similarly, large numbers of these machines are not being recycled. Thus, it is assumed that the majority are stored for later disposition, with the remaining units recycled.

Finally, stored machines are assumed to have been deemed obsolete by the user, or else they would have been re-used. Similarly, they must have been seen to have some relative value—or they would have been recycled or landfilled earlier. Thus, storing is merely an activity that the user does in order to potentially extract future value from obsolete computers. At this point (which happens between seven and eleven years after initial purchase) little chance of effectively reusing this equipment is seen. Most of these machines end up being transplanted to computer recycling firms, yet some are still landfilled. It should be noted that, most likely, this recycled or landfilled decision is more likely made on a volume basis rather than a unit basis. For example, a user may elect to throw away 25 percent of the original system, but recycle the remaining 75 percent. If anything, the percent landfilled is probably lower than this amount. About one-half of the contribution of landfilled computers annually in this model will come from stockpiled computers.

Based on these assumptions, Table 3 summarizes the parameters assigned to variables in the model.

Using this new formulation of the model, a new picture of the computer disposition issue is seen. The updated model predicts over 580 million cumulative PCs will be sold in the U.S. by 2005, but only 72 million computers will be recycled.[1] It predicts that 154 million will be landfilled. Interestingly, these estimates are comparable to the original 1991 study, which depended on highly controversial assumptions and was forecasting 15 years into the future. While the original model was overly simple, it presented a valuable approach.

Table 3: Assumptions used in disposition models.

Parameters	Assumed value
Initial lifetime of a PC	4 years
Obsolete re-used	45%
Obsolete recycled	5%
Obsolete stockpiled	45%
Obsolete landfilled	5%
Lifetime of re-used PC	3 years
Re-used recycled	20%
Re-used stockpiled	70%
Re-used landfilled	10%
Lifetime of stockpiled PC	3 years
Stored recycled	20%
Stored landfilled	80%

[1] A summary of these results and a downloadable Microsoft Excel spreadsheet is available on the Web at <http://gdi.ce.cmu.edu/comprec/>.

5. IMPLICATIONS FOR END-OF-LIFE MANAGEMENT

5.1 Product Take-Back

A more interesting statistic than either the number of computers recycled or landfilled is the number available for product take-back or return. Product take-back is an emerging international paradigm which requires that firms organize methods to reclaim their products at the end of their useful life. Seven European countries have some form of take-back legislation enacted already, and firms are becoming increasingly aware of the possibility that they will be required to comply. In this model, the number available for take-back is defined as those which have either been landfilled or are currently in storage. In 2005, despite increased recycling efforts, 114 million computers will still be available for take-back.

To more completely understand the environmental impacts of personal computer systems, they should be broken down into their major sub-components: motherboards, cases, drives, monitors, and user input devices. This breakdown is basically unchanged since 1991, other than the fact that optical drives (e.g., CD and DVD) are now a standard part of all systems (and the previously mentioned growth of laptops). This is the definition used in subsequent analyses.

It has probably become apparent to the reader that storing obsolete computers in attics or storerooms is inadvisable, since their value will only decrease over time until they are worth only the sum of their raw materials. Further, relatively few ever end up being returned to productive use. The residual value of materials in old electronic equipment soon after production is only 1 to 5 percent of the original cost of the equipment (Matthews 1997). Storing an old computer, instead of quickly reusing or recycling it, is not a profit-maximizing or cost-effective solution. Holding on to an obsolete computer without getting some benefit out of it is akin to holding onto a stock of a company which is known to be going bankrupt—in either case, you end up with nothing, and have full knowledge of this fact ahead of time. Anecdotally, it might be better to sell a four-year-old computer for $200 than to try to sell a seven-year-old computer for $50.

Further complicating the issue of storage is that many computers that are well beyond anyone's definition of "useful life" can only be disassembled for raw materials. Often, choosing this option requires paying a fee to a computer recycler just to comply with local regulations banning disposal. Thus, re-use and recycling need to be more forcefully promoted as options to maximize life cycle values of machines.

5.2 Landfills

Our updated model predicts that 154 million PCs will be landfilled by 2005. In addition, some portion of the 72 million recycled computers will also be landfilled. In this section, the mass of PCs and related material that will be landfilled over time is estimated.

The totals of PC sales include historic specifications of desktop and laptop machines (i.e., they represent average volumes over the 25 years of PCs included in the model). A desktop machine weighs roughly 50 pounds (23 kg) and occupies about 3 cubic feet (0.1 cubic meters) of space. A laptop machine, including accessories, weighs roughly 7.5 pounds (3.5 kg) and occupies 0.14 cubic feet (.01 cubic meters) of space. The fraction of laptops in total PC sales is 20 percent, and growing at an annual rate of 20 percent. Based on personal discussions with several recycling firms, it is estimated that 10 percent of recycled PCs by weight are landfilled.

Compared with a 1991-era computer, these numbers are increasing, mostly because new technologies are being incorporated as standard components and monitors are getting bigger. A computer system sold in 1991 was defined as weighing 30 pounds (14 kg) and occupying 2 cubic feet (0.1 cubic meters) of space. Due to the increasing growth in sales, it is assumed that the typical system is more like the system of today, and use it in calculating landfill requirements.

Using our prediction of 154 million whole PCs being landfilled by 2005, it is further assumed that 80 percent are desktop systems and 20 percent are laptops (conservatively following sales data). Thus, the "weighted average" landfilled machine weighs 42 pounds (19.1 kg) and occupies a volume of 2.4 cubic feet. The 154 million whole PCs landfilled will require 373 million cubic feet of landfill space in the U.S., and the 10 percent of scrap recycled computers will require an additional 17 million cubic feet by 2005, for a total of 390 million cubic feet, or about 3 million tonnes. As a reference point, this volume is roughly one football field of area piled about three kilometers high. Given the 128 million tons of municipal solid waste (MSW) sent to landfills per year (EPA 2002), this amount is relatively small.

Despite its low relative volume in municipal solid waste, the composition of electronic equipment contains toxic substances that could cause environmental damage if they escaped from landfills. This is a potentially much greater concern than finding landfill space for all obsolete PCs, and was a big motivator in the WEEE legislation.

While this model has been based on U.S. data and assumptions, it could be easily used for other countries and regions. The U.S. case is important for several reasons: first, it remains the largest market for PCs in the world; second, most countries are like the U.S.—with no national computer recycling policy in place; and third, it is representative of many other electronic products, like PDAs, mobile phones, etc.

6. LIFE CYCLE ENVIRONMENTAL MANAGEMENT OPPORTUNITIES FOR COMPUTERS

The remainder of this book will discuss and show several approaches to understanding and minimizing the impacts of electronics. Beyond these high-level ideas, there are some more fundamental goals that would aid in reducing the overall life cycle impacts of electronics. If implemented through corporate or government policy, these mechanisms could significantly reduce the volumes of ICT equipment being sent to landfills. These mechanisms are summarized below.

6.1 Design for Environment Programs

The key to successfully improving the environmental quality of any product is to make informed decisions at the design stage. So-called design for environment (DFE) or green design programs maximize use of resources, and also ensure that corporate environmental goals are met in a timely manner. Most major computer manufacturers have environmental design programs in place and publicize their efforts.

6.2 Modular Design and "Upgradeability"

Modular design and upgradeability is intended to alleviate some of the need to constantly upgrade equipment, and thus, to reduce potential waste. Although most computers can be partially upgraded by swapping components like larger drives, few computers exist with appropriate price-performance upgrade paths, and even fewer are able to be traded in for newer models. Apple's PowerMac 7500 series was designed for upgradeability right out of the box. Even the original computer came with the processor on a "daughtercard" attached to the motherboard. Upgrading was as simple as changing daughtercards. Although full upgradeability can never exist due to rapidly changing architectures, an option like this helps to extend the life of computers.

6.3 Component Re-use

In the absence of upgrade paths for most equipment, re-use needs to be seen as a viable option. Most subsystems (e.g., drives, memory, keyboards) are designed to last well beyond the five-year useful life span of the overall computer system. This is an awkward situation, as the whole system sits idle while its subsystems could be re-assigned, yet the purchaser paid more for components that will far outlast the system.

6.4 Materials Selection

One of the more important reasons for concerning ourselves with materials selection is in preventing the use of toxic materials. Generally speaking, it will be impossible to remove all toxics from the design of computers. We consider a few examples of toxic substances in electronic products. While some of these substances are no longer used in new systems, they were in the past, and are listed because they may be in already-landfilled systems. Overall, materials choices fall into three categories: Class I, materials not necessary for operation; Class II, materials necessary for operation and expensive to replace; and Class III, materials necessary but with no easy replacement.

Class I materials choices mentioned in the 1991 study included the lead shielding for CPU cases. Much progress has been made in this area. Manufacturers have switched to cases that use more plastic for shielding, as well as using metals other than lead. Class II materials included pentachlorophenol (PCP) in capacitors, cadmium in batteries, lead solder on circuit boards, and mercury in batteries and switches. Cadmium is still present in rechargeable portable computer batteries, as are traces of mercury, but each in lower quantities than before. Nickel-cadmium (NiCd) technology continues to be the most widespread choice in the industry, although other options such as nickel-metal-hydride (NiMH) exist. Lead solder is still a standard part of circuit board fabrication, but large-scale use of lead-free solder is on the horizon. Finally, Class III materials include phosphorous in monitors, copper-plastic interface cables, and silicon and arsenic in integrated circuits. Little progress has been made in removing these materials from design other than through reduction. Thankfully, projections of widespread use of highly toxic gallium-arsenide (GaAs) silicon wafer technology have proved premature.[6]

An interesting side-effect of the continued use of metals in the design of computers is that they account for a high percentage of the end-of-life value of products. The more metals present (e.g., aluminum), the higher is the reclamation value for a recycling firm. Ironically, as metals are successfully replaced, there will be less incentive for recycling to occur. Metals account for over 70 percent of the residual value of computers (Matthews 1997).

Aside from improvements in materials selection, progress has been made in subsystem recycling. This advance has come in part through legislation, like bans on the landfill disposal of cathode ray tubes (CRTs) and the WEEE legislation in Europe (see Chapter 4). As a result, monitor and CRT recycling has made great advances. This is significant because monitors contribute nearly 50 percent of the mass and volume of computer systems and contain the toxics lead, phosphorus, cadmium, and mercury. Glass, circuit boards, and wiring from both CRTs and standard televisions are reclaimed in great quantities every year (Envirocycle 2002).

Even with these recycling technology advances, there are considerable obstructions to successfully entering the market. A company choosing to reclaim old electronic products is classified as a "hazardous waste handler" by EPA regulations. Improvements need to be made which provide regulatory relief for companies that seek to both reduce landfill waste and improve environmental quality. In the U.S., the Environmental Protection Agency is considering a change in classification for CRTs from a characteristic hazardous waste to a "universal waste." This change in classification would reduce the regulatory barriers for CRT glass recycling.

---

[6] Gallium-arsenide components are seen to be much more toxic per unit of mass than silicon-based devices. While the technology is generally feasible, it is still more expensive than silicon and thus has not been heavily used in mass-market applications.

6.5 Labeling and Materials Recovery

Individual firms have followed the International Standards Organization (ISO) guidelines with respect to labeling parts. IBM has labeled all plastic parts of their computers so that they can be easily identified at end-of-life, using ISO specifications as well as supplier information. Such efforts have, for the most part, caused manufacturers to realize the benefits in reducing the number of materials used (namely, plastics) in their products.

In addition, the electronics industry has made considerable progress in setting up reverse logistic networks and reclamation facilities to extract value from end-of-life equipment. Hewlett-Packard (HP) and IBM both have corporate centers in place to process obsolete electronic equipment.

6.6 Supplier Management

As international environmental quality specifications take hold, electronic firms will need to work in tandem with their suppliers to ensure that the final environmental footprint of the product is in accordance with all regulations. The primary means by which this will occur is by tightening the information linkages between suppliers and manufacturers. Original equipment manufacturers (OEMs) like Siemens and IBM are already querying subsystem manufacturers like Intel and Quantum regarding the materials, chemicals, and processes used. Subsystem manufacturers are now querying their own suppliers regarding the same issues. As these relationships grow, significant environmental benefits should be seen as the entire industry seeks to reduce environmental impacts.

6.7 Recycling Promotion

As mentioned above, used electronic equipment loses value quickly. Generally speaking, computers depreciate at a rate of 40 percent per year. Manufacturers are aware of this and realize the environmental burden of their products. In the U.S., the Electronic Industries Alliance (EIA) has begun an educational campaign to inform users of the environmental impacts of electronic products, and is starting several pilot programs to more effectively determine optimal end-of-life management paths.

Changes in government procurement and disposition procedures would be helpful as well. Currently, there are restrictions in government procurement guidelines preventing the purchase of recycled products. Also, many products purchased by the government cannot be donated or otherwise re-used, requiring long-term storage of obsolete equipment.

6.8 Resource Recovery and Product Take-back

Although several major manufacturers have the ability to reclaim materials from products, most such reclamation is being done by third-party groups. Although this inevitably meets the goal of preventing disposal, manufacturers thus have little incentive to try to design products that use non-virgin materials. Resource recovery programs are critical to closing the loop towards having a steady supply of non-virgin materials for manufacturers.

Large firms will sometimes negotiate take-back pricing for old equipment (in the $50 range), but such a decision soon becomes moot, as donation firms will give $100 or more tax credits for the same equipment. This fact underlies the economics of the situation—that personal computers are relatively expensive to completely recycle and there are small gains, whereas mainframe computers can be disassembled easily and produce high returns. Note that this research has not even considered the potential environmental impacts or feasibility of software package recycling.

Firms with the logistics available to take back products should capitalize on this and incorporate outreach programs to fill capacity. Many firms donate old equipment to charity—Why not promote recycling by launching "computer drive-off" programs, where empty trailers are left at schools to be filled with old equipment? In exchange for filling a certain number of trailers, the manufacturer provides "new" equipment to the school. Consumers have shown a remarkable willingness to provide computers for schools via grocery store receipt collection. Such programs could further inform the public about the incentives for recycling old equipment.

Several large national electronics retailers in the U.S. have experimented with allowing customers to bring in obsolete equipment for recycling. Programs like this have a double benefit of improving recycling rates and also increase business to these companies acting in the social interest.

7. CONCLUSIONS

The growth rate of information and communication technology industries has been dramatic over the past decade. Consumers are buying more portable computers and devices than ever before. ICT products, especially computers, create extensive environmental damage in numerous forms throughout their life cycles. There are large uses and management of toxic substances needed to produce electronic products and their components; the batteries and electricity needed to power them over their rapidly obsolescent lives also cause environmental impacts. A trend towards even smaller, harder-to-recycle products will only intensify the environmental problems.

At end-of-life, many computers are simply stored by users, leading to costly disposal issues. Good data on sales of devices exist, but estimates of product life spans, recycling rates, and other pathways are not systematically maintained. Thus, concerns about these effects need to be addressed by creating models and estimates of equipment volumes. Currently, reclamation businesses for electronic products are flourishing from a constant stream of obsolete products, thereby reducing the number of units going to landfills and increasing the number available for take-back. The quantities of recycled electronic products are still small, however, compared to the stocks of equipment outstanding. As mandatory take-back policies go into effect around the world, such reclamation activities will become strategically valuable as scale efficiencies are gained and more public awareness of the problem is achieved.

ACKNOWLEDGMENTS

H.S. Matthews acknowledges the support of the AT&T Industrial Ecology Faculty Fellowship program.

REFERENCES

Berkhout, Frans and Julia Hertin. 2001. Impacts of information and communication technologies on environmental sustainability: Speculations and evidence. Draft Report to OECD Directorate for Science Technology and Industry. Brighton: Organization for Economic Cooperation and Development.

Commission of the European Communities (CEC). 2000. Draft proposal for a European Parliament and Council directive on waste electrical and electronic equipment. Brussels: Commission of the European Communities.

Cellular Telephone Industry Association (CTIA). 2001. *Semi-annual wireless survey*. March 2001.

Curran, M. A. 1996. *Environmental life-cycle assessment*. New York: McGraw Hill.

Dataquest Corporation (now Gartner). 1996–2003. Press releases, as well as secondary publication in *Statistical abstract of the United States, 1991–1995 editions*.

Environmental Protection Agency (EPA). 2000. Life cycle assessment brief. July 2000.

———. 2002. *Municipal solid waste in the United States: 2000 facts and figures*. EPA530-S-02-001, June 2002. Washington, D.C.: United States Environmental Protection Agency, Office of Solid Waste and Emergency Response.

Envirocycle, Inc. 2002. Three companies achieve breakthrough in CRT recycling. Press release, 18 June 2002.

Fava, J., R. Denison, B. Jones, M. A. Curran, B. Vigon, S. Selke and J. Barnum, eds. 1994. *A technical framework for life cycle assessment.* Pensacola: The Society of Environmental Toxicology and Chemistry.

Hocking, M. B. 1991. Paper versus polystyrene: A complex choice. *Science* 251 (1 February): 504–505.

Industrial Council for Electronic Equipment Recycling (ICER). 2000. *ICER status report on waste from electrical and electronic equipment*. London: Industrial Council for Electronic Equipment Recycling.

Internet Industry Almanac. 2002. USA tops 160M Internet users. Press release, 16 December 2002. <http://www.c-i-a.com/pr1202.htm> (25 August 2001).

International Telecommunications Union (ITU). 2001. Key global telecom indicators for the world telecommunication service sector. <http://www.itu.int/ITU-D/ict/statistics/at_glance/KeyTelecom99.html> (26 May 2003).

Matthews, H. Scott. 1997. Unraveling the environmental product design paradox. Proceedings of the 1997 IEEE Symposium on Electronics and the Environment, May 1997, at San Francisco, CA.

Matthews, H. Scott, Francis C. McMichael, Chris T. Hendrickson and Deanna Hart. 1997. Disposition and end-of-life options for personal computers. Carnegie Mellon University Green Design Initiative technical report 97–10. Pittsburgh, PA: Carnegie Mellon University.

National Safety Council. 1999. *Electronic product recovery and recycling baseline report*. Washington, DC: National Safety Council.

Organization for Economic Cooperation and Development (OECD). 2000. *Measuring the ICT sector*. Paris: Organization for Economic Cooperation and Development.

Rosenblum, J., A. Horvath and C. Hendrickson. 2000. Environmental implications of service industries. *Environmental Science & Technology* 34(22): 4669–4676.

Schreyer, P. 2000. The impact of information and communication technology on output growth. OECD STI working paper, 2000/2. Organization for Economic Cooperation and Development.

SEMATECH. 2000. International Sematech audit shows marked decrease in ultra pure water use. Press release. <http://www.sematech.org:80/public/news/releases/wateruse.htm> (15 August 2001).

Townsend, T. G., S. Musson, Y. C. Jang and I. H. Chung. 1999. Characterization of lead leachability from cathode ray tubes using the toxicity characteristic leaching procedure. Report #99-5, December 1999. Florida Center For Solid And Hazardous Waste Management.

U.S. Office of Technology Assessment. 1992. *Green products by design: Choices for a cleaner environment* (OTA-E-541). Washington, DC: U.S. Government Printing Office, October 1992.

Vigon, B. W., D. A. Tolle, B. W. Cornaby, H. C. Latham, C. L. Harrison, T. L. Boguski, R. G. Hunt and J. D. Sellers. 1993. *Life cycle assessment: Inventory guidelines and principles* (EPA 600/R-92/245). Washington, DC: U.S. Environmental Protection Agency.

Chapter 3

ENVIRONMENTAL IMPACTS IN THE PRODUCTION OF PERSONAL COMPUTERS

Eric Williams
United Nations University, Japan

1. INTRODUCTION

The environmental impacts associated with a product include those that occur in the production processes to make it. These impacts occur not only at the final stage of assembling parts, but also in the production of parts and their constituent materials. Environmental analyses often put emphasis on impacts from producing the raw materials to make products, such as those arising from mining, metal smelting, and refining petroleum. From this perspective, a personal computer (PC) does not make much of an impression. A PC is much smaller than a car, for instance, and being a high-tech product, was presumably produced by a "clean" set of processes. The objective of this chapter is to examine this issue more carefully, starting with the basic question: What are the types and extent of the environmental impacts associated with making a computer? This examination will also consider hazardous materials that may leach from end-of-life computers in landfills. While this is ultimately a disposal issue, the materials are put in the computer at the production stage so should be considered here.

The environmental impact of manufacturing computers is a difficult and complex question to address. One problem is the lack of reliable data detailing the processes in the production chain. This problem exists to a certain extent for all products, but is exacerbated in high-tech industries; processes evolve rapidly and technical know-how is a key factor in firm competitiveness. Releasing information about processes is thus often perceived as a business risk. Another obstacle is more fundamental: How should one interpret "environmental impact"? One product may require considerable energy to produce and thus give rise to greenhouse gas emissions, but another product may result in larger emissions of acids to water systems. Which is "worse"? This is ultimately a subjective question that requires input from society on how to prioritize environmental

Computers and the Environment: Understanding and Managing Their Impacts
Edited by Ruediger Kuehr and Eric Williams, pages 41–72.

problems. In addition, there may be environmental benefits associated with the use of a product. For instance, if communicating via the Internet replaces a degree of automobile use (by telecommuting to work, for instance), this could reduce energy consumption and transport-related emissions. There is much activity in the environmental research community devoted to trying to address these questions, but it is not the intention here to delve into such complexities. Rather, this chapter has two purposes: one is to survey the main environmental issues associated with producing computers in general terms; the other is to describe the flow of materials in manufacturing processes and undertake an analysis of the total amounts of energy, chemicals, and water used to make a desktop computer. The latter is a rather technical task and the detailed discussion of processes, data, and the calculations may be of most interest to specialists in life cycle assessment.[2] Removing these details would certainly make the chapter more readable and a great deal shorter, but a degree of readability will be sacrificed for transparency, because until now no published environmental analysis of computers has laid out its underlying data and assumptions. Still, there is much material here of interest to the non-specialist (the sections on materials analysis can be skipped without loss of continuity).

2. PRODUCING A COMPUTER

A discussion of the network of processes involved in making a computer is an appropriate starting point for the analysis. Figure 1 shows a simplified pictorial version of the network of processes: The real system is far more complex, involving hundreds of processes and companies. The network is decentralized and global, with computer manufacturers increasingly focusing on design and assembly only, and components traded between firms on the international market.

Surveying the environmental issues associated with an entire production network is challenging, to say the least. This chapter will approach the task by isolating particular industrial sectors thought to be of particular environmental relevance for relatively detailed discussion (such as microchip fabrication and circuit board production), while other processes will be described in an aggregate way. Note that there is no specific mention in Figure 1 of the production of disk drives and various non-semiconductor components such as capacitors; the analysis will only consider them in terms of the bulk materials they contain.

[2] Life cycle assessment (LCA) is a technique to quantitatively estimate the environmental burden associated with the total life cycle (production, use, disposal) of a given product or service.

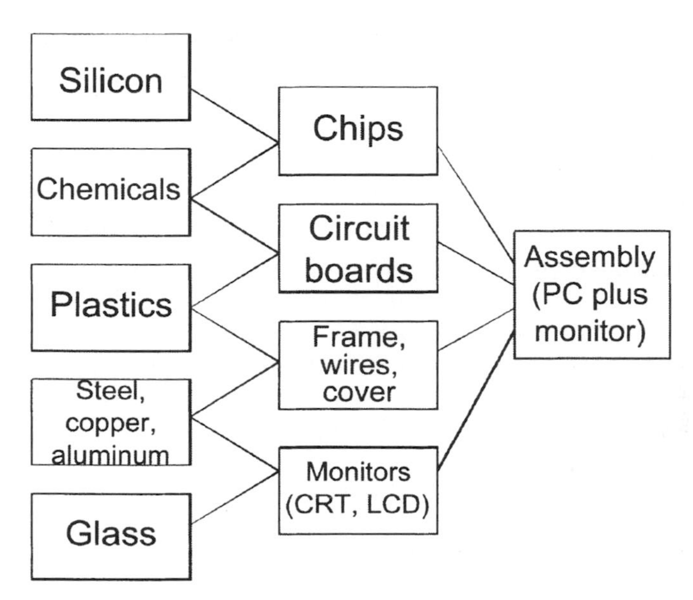

Figure 1: Production network for computers.

The environmental issues associated with an industrial sector are described in this chapter from two perspectives: materials analysis and a qualitative survey of key impacts. Materials analysis, often used in the fields of industrial ecology and life cycle assessment, describes a process in terms of the masses of inputs of raw materials as well as outputs of products and emissions of wastes (Ayres and Ayres 2002; Curran 1996). The actual environmental impact of a sector is more complex than its material flow, but the description gained from materials analysis is useful in gauging the possible scale and types of environmental impacts that can occur. The qualitative survey covers particular environmental issues facing PC-related industries, such as concerns over occupational safety and health.

3. SURVEY OF MANUFACTURING PROCESSES AND THEIR ENVIRONMENTAL IMPACTS

Based on the above sketch of the production network, this section describes in more detail the economic status, technology, materials/energy use, and environmental issues associated with particular manufacturing processes. Six aspects of the manufacturing network are considered:

1. Fabrication of microchips,
2. Manufacture of printed circuit boards,
3. Manufacture of cathode ray tube (CRT) monitors,
4. Manufacture of liquid crystal display (LCD) monitors,

5. Production of bulk materials in computers and monitors (steel, plastic, etc.),
6. Production of specialized chemicals and materials for electronics manufacturing.

3.1 Microchips

3.1.1 The Global Microchip Industry and its Production Processes

A computer is essentially a vehicle for microchips (or more generally, semiconductor devices) to do their work. Although the semiconductor industry has been in a slump since the collapse of the high-tech "bubble" of the 1990s, viewed over the long term it has grown rapidly over the last few decades and has become a sector with global economic clout. Average growth in economic value from 1970 to 1999 was 16 percent per annum, and the total economic value of the industry globally in 2001 was U.S.$139 billion (SIA 2002). Shipments of the U.S. automobile industry, by comparison, are valued at around $200 billion. Roughly 50 percent of chips produced are for use in computers, 19 percent for telecommunications equipment, 16 percent for consumer electronics, 10 percent for industry, and 6 percent for automobiles and aerospace applications (Stuart 1997). The geographical distribution of production shares in 2001 was East Asia (mainly Korea, Taiwan, and China) with 29 percent; the United States, 27 percent; Japan, 24 percent; and Europe, 22 percent (SIA 2002). The industry is highly global in nature and largely under the control of multinationals headquartered in the United States and Japan.

A process called wafer fabrication lies at the center of microchip production. This involves building a complex patterned stack of layers on top of a silicon wafer. It is this pattern that becomes the network of transistors and diodes that give a chip its function. Fabrication technology is often described in terms of the smallest size of features in the pattern (i.e., distance between two transistors on a chip); a typical size in recent years is 0.18 μm (0.00018 millimeters). Improvements in chip performance over the years are mainly due to advances that allow this pattern size to be made ever smaller. A denser pattern implies that one can fit more transistors (i.e., more functionality) on a chip, and performance becomes faster due to shorter travel distances between components. The famous Moore's Law reflects the fact that the rapid pace of technological progress is in some sense constant; the number of transistors that can be put on one chip has doubled every 18 to 24 months for at least three decades.

The process of making chips is complex; only the basics are described here (Van Zant 1997). In general terms, chip fabrication involves a sequence of layering, doping/oxidation, and patterning processes. *Layering* involves laying down a thin layer (0.5–5 μm) of a desired material, usually silicon or a metal such as aluminum. *Doping* involves adding small amounts of an element, such as boron or phosphorus, that changes the electrical properties of the silicon layer, and *oxidation* changes a semiconducting silicon layer to an insulating silicon dioxide one. *Patterning* is the carving of a dense, maze-like set of furrows into a layer. It is based on photolithography, a process rather similar to conventional photography. A layer of photoresist (a substance that becomes either more or less dissolvable with respect to a given solvent when exposed to light) is put on top of the wafer. Next, ultraviolet light is shone onto the photoresist layer through a photomask, which is a glass sheet with the maze-like pattern intended for the chip etched onto the surface. This pattern is transferred to the wafer in the form of a pattern of exposed and unexposed regions. Next, the pattern is developed, applying special solvents that specifically dissolve only exposed (or unexposed) photoresist. Up to this stage, the silicon layer underneath the photoresist has not been altered; this is done in the next step, known as etching. Etching is either done using solvents that attack the layer to be patterned, but not the photoresist, or by bombardment with particles that chip away at both the layer to be carved and the photoresist layer. The final step is stripping, the removal of the now-unneeded photoresist, through a bath in a solvent that dissolves photoresist but not other layers on the wafer. Between process steps, wafers often need to be cleaned with ultra-pure water, acids, or other agents. The desired stack of layers is made by repeated application of the cycle (layering, doping/oxidation, and patterning), using different masks for each layer.

The end result of the fabrication process is a wafer with the patterned layers for a number of chips. The wafer is sliced up into individual functional pieces, called dies, and then electrically connected to a frame that includes the chip's external "legs," and then encased in an epoxy or ceramic package to protect the functional parts inside.

3.1.2 Materials and Energy Used in Microchip Production

The chip fabrication process uses considerable amounts of chemicals, water, and energy. Quantitative data for inputs of silicon, aggregate chemicals, fossil fuels, and water are shown in Table 1. Some tens to hundreds of chemicals are used; detailed breakdowns showing use of individual chemicals can be found in previous work of the author and collaborators (Williams 2000; Williams et al. 2002). The third column of Table 1 refers to the amounts required to produce one 32-megabyte (MB) DRAM memory chip (Williams et al. 2002). The next column is an estimation of material use by the global semiconductor industry, obtained by simply multiplying the amount required per unit area of silicon by the 1998 silicon wafer consumption of the worldwide industry, totaling 28.4 billion square centimeters (cm^2)[3] (Yamauchi 2000). The last column shows estimates of amounts required to produce the chips contained in one average computer. These estimates were obtained by first assuming that material use is proportional to the economic value of chips used. The computer industry uses 49 percent of world semiconductor production (by economic value), and the vast majority of these were used in the 90 million personal computers produced in 1998 (Crothers 1999). The "embodied fossil fuels" figure in the energy data reflects both direct

[3] As the wafer fabrication process builds layers on the surface of a wafer, using area as a measure of wafers is more useful than using mass.

consumption in processes and indirect use to produce the electricity purchased for chip production. Although the amount of fossil fuels needed to generate electricity varies from nation to nation according to different mix of energy sources used (e.g., more hydropower and less coal implies reduced fossil fuel use), throughout this analysis the global average is used (IEA 2002). It is calculated by multiplying electricity use by 320 grams of fossil fuel per kilowatt-hour (kWh),[4] based on the global average mix of electricity-generating technologies (coal, hydropower, nuclear, etc.) and assuming that the average energy density of directly consumed fossil fuels is 40 megajoules per kilogram (MJ/kg).[5]

Although these estimates are undoubtedly crude, they provide a reasonable guess at the order of magnitude. The main point to note is that the materials used per chip and per computer are considerable, which suggests that the environmental impacts of the industry are also significant. For microchip production, the total fossil fuels (970 grams) and chemicals (72 grams) used are 500 times the weight of a two-gram chip. By comparison, a 1,000-kilogram automobile requires only twice its weight in fossil fuels for the entire production chain (MacLean and Lave 1998).

When confronted with such figures, representatives from the semiconductor industry often respond that the results are based on data that is several years old (inevitable if using macro statistics) and that much improvement has been made since then. While it is true that the industry has made progress in reducing materials and energy use, efficiency improvements have been slower than the growth of the industry; thus total materials and energy use continues to climb rapidly. For instance, given a backdrop of the 16 percent average economic growth of the global semiconductor sector, use of electricity by the U.S. semiconductor industry increases on average 7 percent per year and chemical use by 8 percent per year (U.S. Census Bureau 2001; Kanegsberg and Van Arnum 1999).

3.1.3 Environmental Impacts of Semiconductor Production

While use of fossil fuels is closely correlated with impacts on climate change and other issues such as acid rain, the amount of aggregate chemical consumption and waste provides but a crude indicator of potential environmental damage. The current understanding of environmental impacts associated with chemical use in semiconductor manufacture is summarized below.

3.1.3.1 Chemical Emissions

A wide variety of chemicals are used in semiconductor fabrication, many of them toxic, hence the potential impacts of emissions on air, water, and soil are major environmental concerns. Emissions of chemicals can potentially harm the environment in a number of ways. Excessive discharge of acids or alkaline solutions into water

---

[4] One kilowatt-hour (kWh) = 3,600 MJ.

[5] A megajoule (MJ) is the standard energy unit in the metric system.

systems can affect microscopic, plant, and animal life. Many organic and inorganic chemicals are acutely toxic to wildlife and humans given sufficient exposure. Of particular concern here are the possible long-term health effects of chemical exposure on cancer rates and reproductive health, which are often difficult to determine.

Table 1: Materials and energy used in manufacturing microchips

Material	**Description**	**Amount per memory chip**	**Annual use by industry worldwide (weights in metric)**	**Amount used to make chips in one computer**
Silicon wafer		0.25 grams	4,400 tons	0.025 kg
	Dopants	0.016 grams	280 tons	0.002 kg
	Photolithography	22 grams	390,000 tons	2.2 kg
Chemicals	Etchants	0.37 grams	6,600 tons	0.037 kg
	Acids/bases	50 grams	890,000 tons	4.9 kg
	Total chemicals	72 grams	1.3 million tons	7.1 kg
Elemental gases	N_2, O_2, H_2, He, Ar	700 grams	12 million tons	69 kg
	Electricity	2.9 kWh	52 billion kWh	281 kWh
Energy	Direct fossil fuels	1.6 MJ	28 billion MJ	155 MJ
	Embodied fossil fuel	970 grams	17 million tons	94 kg
Water		32 liters	570 billion liters	310 liters

Source: Adapted from Williams et al. (2002).

There are a number of cases from the 1980s where wastes from the semiconductor industry contaminated water supplies. The main source of these emissions was from accidental leakage of chemicals from storage tanks. The earliest and most famous case occurred in Silicon Valley, in which 15,000 liters of 1,1,1-trichloroethane (an organic solvent) leaked from the storage drums of Fairchild Semiconductor from 1977 to 1981, entering local water supplies (Yoshida 2002). An epidemiological study carried out by local authorities indicated a noticeable rise in miscarriages and birth defects in the affected areas (Swan et al. 1985). A separate study showed a correlation between exposure to 1,1,1-trichloroethane and cardiac abnormalities in rats (Dapson et al. 1984). Citizens filed a class-action lawsuit, and in 1986, Fairchild, IBM, and the local water company paid an undisclosed multimillion-dollar settlement to 530 residents. Contamination cases (without the lawsuits) have also cropped up in Japan. Groundwater near the electronics plants of Toshiba in Kimutsu and Mitsubishi in Kumamoto were found to contain concentrations of tetrachloroethylene and trichloroethylene far exceeding regulated standards (*Mainichi* 1998, *Bernama* 1998). There is as yet no evidence, however, that these leakages resulted in damage to human health.

The industry has made significant efforts to reduce environmental impacts through chemical substitution, reduction of use, and improved treatment and storage technologies. At least in the United States, obvious contamination cases like those in the 1980s did not re-appear in the news of the 1990s—a sign that the situation has probably improved. As the amount of publicly available data is scarce, however, it is difficult to accurately judge the current situation. It is also worth considering that although production of semiconductors was until recent years dominated by the industrialized world, production in East Asia, especially in Taiwan and China, has dramatically increased. Production is, to a large degree, under the control of multinationals that, hopefully, implement waste treatment systems on a comparable level as practiced in the industrialized world. There is almost no information available, however, regarding the environmental practices of subsidiaries of multinationals or local firms in the industrializing world.

3.1.3.2 Health of Semiconductor Workers

The occupational health and safety of workers on fabrication lines is of particular concern. Because semiconductor production is a light manufacturing industry, the worker illness and injuries that occur day-to-day are not as severe as in heavy industries. The main health concern for the industry is whether long-term chemical exposure increases rates of birth defects and cancer. The extent of risk to workers is still unclear; connections between long-term exposure and illness are notoriously difficult to prove (or disprove), and this is especially true for the semiconductor industry, due to the complexity of its chemical mixes and frequent changes in processes. Epidemiological evidence points to potential negative effects on reproductive health: three studies indicated that female fabrication workers displayed an elevated incidence of miscarriages (Pastides et al. 1988; Correra et al. 1996; Schenker et al. 1995). This increased miscarriage rate was correlated, though not conclusively, with exposure to ethylene glycol compounds (used as a photoresist solvent). Many firms reportedly stopped using these chemicals as a result of these studies.

Elevated miscarriage rates are considered a warning signal of possible elevated rates of birth defects and cancers. IBM and National Semiconductor have been facing a barrage of lawsuits from former workers claiming that chemical exposure led to birth defects or cancer. Most cases are still pending, but those completed were settled out of court for undisclosed compensation and no admission of guilt (Fisher 2001; Hellwege 2002). There is no scientific evidence either supporting or denying workers' claims. Only one publicly-available epidemiological study investigating birth defects and cancers in the semiconductor industry currently exists. This study, carried out by health authorities in the United Kingdom, found that cancer rates for workers at a Scottish plant were comparable to the average population, though there were noticeable distinctions in the patterns of cancer that developed. Plant workers showed higher incidences of lung, stomach, breast, and brain cancer. This work has been criticized, however, due to its small sample size (71 deaths) and lack of distinction made between different work duties at the plant (Bailar et al. 2000). There has been a long-standing need for a serious study of cancer and birth defects in the semiconductor industry, but so far firms and governments have shown little initiative to carry out such work (Bailar et al. 2002).

3.1.3.3 Energy and Water Use

The data in Table 1 suggest that the semiconductor industry is energy-intensive. For a macro-view on the issue, note that electricity use of the Japanese semiconductor sector was reported at 5.4 billion kilowatt-hours in 1995, which represents 0.55 percent of national electricity demand. If energy use continues to grow as rapidly as it has in recent years, this share will reach 1.7 percent by 2015. This is a significant but not overwhelming amount for a single production sector. It is important to include the use of perfluorocarbons (PFCs) when considering the contribution of the sector to global warming. PFCs (such as CF_4 and C_2F_6) are used in etching processes and are potent greenhouse gases. According to 1993 estimates by SEMATECH, an international research consortium of the semiconductor industry, the global warming impact (in carbon dioxide equivalent) of PFC emissions in manufacturing is of a similar order of magnitude to the global warming impact of electricity use by the industry (Williams 2000). The industry has been making efforts to phase out PFC use and, while thus far un-quantified, it is plausible that emissions are declining.

The consumption of water in semiconductor fabrication is also significant. The extent to which this places burdens on water supplies depends on local conditions. A recent drought in Taiwan pitted high-tech firms and farmers against one another in a competition for limited water resources (Yu-Tzu 2002).

3.2 Printed Circuit Boards

3.2.1 The Circuit Board Industry and its Production Processes

A printed circuit board (PCB) (also called a printed wiring board [PWB]), the "green rectangular board with chips attached," has the essential role of electrically connecting microchips and other electrical components. In essence, a PCB is a maze of copper conducting paths stacked between insulating layers, with built-in sockets to receive components. In order to connect components together in more complex patterns, circuit boards are often stacked to make multi-layer boards with ten or even more layers.

The economic value of global production of PCBs in 2000 was $42.7 billion, with an average 9.5 percent growth per annum over the period 1996 to 2000. The circuit board industry is global and competitive, with low margins. Japan recently took the lead in global production, with a 2000 market share of 27 percent, followed by the United States (25 percent), Taiwan (11 percent), China (9 percent), and South Korea (5 percent) (Circuits Assembly 2001).

Production of a PCB begins with a board design that connects a set of components together to achieve the desired function (EPA 1995). The physical base of a board is a uniform sandwich of insulator (usually glass-woven epoxy) and thin copper layer(s). The objective of the manufacturing process is to carve a maze of interconnections on the copper layer(s) and drill small holes for mounting components and stacking boards. The pattern in the copper layer is carved via photolithography and resembles semiconductor fabrication. The photoresist is usually in the form of a dry film laid mechanically on top of the board. The pattern is exposed by shining light through a mask (or negative in photographic terms). Exposed (or unexposed) regions are developed by soaking the board in a substance such as sodium or a potassium carbonate solution. Etching copper is usually done using acidic cupric chloride ($CuCl_2$ and HCl) or alkaline ammoniac ($Cu[NH_3]_4$). Stripping the remaining photoresist is often done with hot potassium hydroxide. The holes must also be coated with a conductor in order to connect the different layers. This is usually done via a process called electroless copper, which involves soaking the plate in a solution with chemicals such as formaldehyde, caustic soda, and copper salt.

The result of this process is one layer of patterned board containing holes. To make a multi-layered board, the process is repeated to make each slice separately, and the final product is made by stacking the layers and melding them together via heat and pressure. In order to attach components, tin-iron solder is screen-printed onto the mounting locations. Components are mounted and the solder melted in a specialized oven. The result is a completed board with components attached.

3.2.2 Materials, Energy Use and Emissions in Circuit Board Production

It is clear from the above discussion of production technology that circuit board fabrication is chemicals intensive. Quantitative information on materials use and emissions of the sector is scarce, however. Table 2 shows data from the Electronics Industry Association of Japan (EIAJ) on aggregate material inputs and outputs in 1995 for the Japanese national industry (EIAJ 1997). The figures for inputs do not include purchased chemicals, although they are partially reflected in the output figures. Starting with the data for Japan, the total use and emissions for the global PCB industry was estimated by simply multiplying the Japan figure by four, reflecting Japan's share in the economic value of global production. The amount of use and emissions to produce the PCBs contained in one computer are estimated by first calculating the intensity to produce 1 kilogram of circuit boards, and then multiplying by 1.85 kilograms to reflect the weight of the circuit board contained in a desktop computer and CRT monitor.[6] The use and emissions per kilogram were calculated using data that put the total weight of Japanese PCB production at 88,600 metric tons (EIAJ 1997). In Table 2 and elsewhere, embodied fossil fuels are calculated from direct fossil and electricity use, as in Table 1.

The quantities of waste acids and alkali per computer, 9.4 and 4.2 kilograms, respectively, are significant—even larger than corresponding figures for

[6] The 1.85 kg figure was obtained by removal and measurement of circuit boards from a Micron Pentium 100 MHz desktop system with a 17-inch monitor.

semiconductor components in a computer. Water use is significant, while energy use is relatively small.

Table 2: Materials and energy use in and emissions from circuit board production (data for Japan; estimates for global industry and per computer).

Inputs	1995 Japan use/ emissions[7]		1995 global use/ emissions		Use/emissions per desktop system	
Blank boards	73,318	tons	293,000	tons	1.70	kg
Resin etchants	2,789	tons	11,200	tons	0.06	kg
Solder	3,188	tons	12,800	tons	0.07	kg
Copper	8,573	tons	34,300	tons	0.20	kg
Aluminum	1,650	tons	6,600	tons	0.04	kg
Plastic	6,265	tons	25,100	tons	0.14	kg
Electricity	1.17 billion	kWh	4.67 billion	kWh	27	kWh
Fossil fuels	244 million	liters oil	975 million	liters oil	5.6	liters oil
Embodied fossil	0.596 million	tons	2.38 million	tons	14	kg
Water	60.2 billion	liters	241 billion	liters	780	liters
Outputs						
Waste acid	408600	tons	1.63 million	tons	9.4	kg
Waste alkali	181234	tons	725000	tons	4.2	kg
Sludge	19771	tons	79000	tons	0.45	kg
Waste plastic	12522	tons	50100	tons	0.29	kg
Other	36906	tons	148000	tons	0.84	kg
Total	659034	tons	2.64 million	tons	15	kg

Source: EIAJ (1997).

3.2.3 Environmental Impacts of Circuit Board Production

3.2.3.1 Generic Issues Concerning Chemical Emissions

The generic aspects of the environmental impacts of chemical emissions described in the section on the semiconductor industry apply here as well. Particular issues for the PCB industry are the large volumes of metals, copper, lead, silver, tin, and chromium that must be managed, or else they can affect the biological components of public wastewater treatment facilities. The electroless copper plating process emits significant quantities of formaldehyde, one the most hazardous chemicals used in PCB manufacturing (EPA 1995).

[7] The decrease in significant figures from left to right is intentional. The first column reporting Japanese emissions is copied from the source (EIAJ 1997).

3.2.3.2 Brominated Flame Retardants

Much concern has arisen in recent years about the possible health effects on humans and animals caused by brominated flame retardants. Compounds such as polybrominated diphenyl ether (PBDE) are routinely added to circuit boards and other goods (such as plastics in computer cabinets, carpets, and upholstery) in order to reduce flammability. These substances are bio-accumulative (stored in fatty tissues and thus magnified as they move up the food chain) and often likened to the notorious polychlorinated biphenyls (the other "PCBs"). The possible threats to human health posed by these compounds splashed onto center stage with the publication of a study in 1998 that showed rapidly rising levels of PBDEs in human breast milk in Sweden (Darnerud et al. 1998). Later studies showed levels in breast milk are also on the rise in North America, doubling every five years (She et al. 2000; Ryan and Patry 2000). Toxicology studies are underway, but already there is evidence that exposure can be linked to thyroid hormone disruption, neuro-developmental deficits, and possibly, cancer. The minimum concentrations needed to induce health effects in humans are not well understood. One well-known study comes to the preliminary conclusion that the minimum level of intake that induces negative health effects is one milligram (mg) per kilogram a day, one-millionth less than the human intake levels measured in Sweden (Darnerud et al. 2001). They stress, however, that there are important data gaps regarding carcinogenicity and effects on reproductive systems. Exposure routes are also poorly understood. Food consumption is thought to be a key route (Wenning 2002), though there is evidence that workers engaged in the demanufacturing of computers showed elevated levels of PBDE and other flame retardant compounds (Jakobsson et al. 2002). The European Union passed, as part of its Restriction on Hazardous Substances (RoHS) Directive, a ban on use of PBDE and polybrominated biphenyls (PBB) by 2006 (European Parliament 2003).

3.2.3.3 Lead in Solder

Legislation recently passed by the European Union (E.U.) will ban the use of lead in electronic products (including circuit boards) starting in 2006 (European Parliament 2003). The rationale for the ban is mainly based on concern that lead will leach from circuit boards in landfills and contaminate water supplies. The experimental basis for this concern is provided by research results showing that ground and shredded circuit boards in water routinely leach amounts of lead exceeding environmental standards (Yang 1993; Townsend et al. 2001).[8] The ban has aroused international debate; some question its environmental benefits and equity with regard to trade relations. China, for example, will no longer be able to export electronic equipment with lead-containing solder to the European Union.

[8] For example, the toxic characteristic leaching potential (TCLP) test shows shredded circuit boards leaching 175 to 205 milligrams per liter (mg/l), much higher than the EPA-regulated maximum of 5.0 mg/l for non-residential landfill waste.

The ban raises important questions on the relationships between scientific uncertainty, environmental regulation, and international trade. All products pose a certain degree of environmental risk associated with their use and disposal. How large does this risk need to be to justify a ban on use of some substance? How is this risk to be evaluated? In what cases can a ban be chosen as a response instead of making recycling mandatory? With today's integration of the global economy, when can a single region justify the choice to a ban on use, when such a ban affects trade agreements? In a similar context, the case of banning chlorofluorocarbons (CFCs), such as Freon, was clearer; the risk to the planet's ozone layer was obviously far higher than that posed by alternative refrigerants, and there was international consensus to ban production via the Montreal Protocol. The situation is different for lead solder in electronics. While circuit boards usually fail a laboratory leaching test, there is no evidence yet that the macro-level risk is significant. Also, are the available lead-free alternatives true solutions to the suspected problem? Both the absolute level of risk of lead solder and the comparative risk of lead-free alternatives are poorly understood. No analyses have yet addressed the basic question of whether current and projected leaching of lead from electronic equipment in landfills is likely to result in contamination of water supplies. One study reports that the environmental impacts of lead solder are worse than its alternatives (Griese et al. 2000), and another asserts that both have impacts and that products made with lead-free solder are not necessarily more environmentally-friendly than leaded ones (Turbini et al. 2000). These are preliminary assessments; the jury is still out on the issue. Also, finding that a substitute material is *relatively* environmentally friendly does not address the question of whether the *absolute* risk is significant or not.

There is dissent internationally and within the European Union regarding the ban on the use of lead in electronic products. The United States has mounted opposition to the ban. Turbini and collaborators point out that lead in circuit boards represents less than 4.4 percent of total lead found in landfills and that a study of leaching from 146 U.S. landfill sites found only two cases of measurable leakage of lead into groundwater, both due to large quantities of lead-containing industrial waste being deposited there (NUS 1988). A leaching study by a U.S. electronics firm found that lead-free alternatives bearing silver or antimony failed national TCLP standards for those elements (Smith and Swanger 1999). There are also estimates that economic costs of switching to lead-free solder will be high and may offer no environmental benefit (Miller 2002).

The positive and negative environmental and economic consequences of the E.U. lead ban remain uncertain. The legislation was apparently passed in an attempt to apply the "precautionary principle," the idea that action to reduce environmental risks should be taken without waiting for conclusive scientific evidence. There is certainly a strong argument to apply the principle in cases where science cannot definitively answer certain questions, such as in the case of climate change. On the other hand, it can also be misapplied if used to justify literally any action by isolating one possibly negative factor. It is reasonable to assert that a necessary precondition for applying the precautionary principle is a reasonable attempt to reduce uncertainty within the limits of contemporary science. Given the lack of studies to assess risk, this attempt has yet to be made for the case of lead in circuit boards.

3.3 Displays: Cathode Ray Tubes (CRT)

3.3.1 The Global CRT Industry and its Manufacturing Processes

A cathode ray tube (CRT) is a display based on a technology wherein an image is created by a controlled bombardment of electrons onto a flat plate coated with phosphorescent substances. CRTs gained their first introduction to the mass markets of the industrialized world with television in the 1950s and 1960s, followed by computer monitors in the 1980s and 1990s. The global CRT monitor industry had an economic scale of $19.5 billion in 2001, producing 108 million units (Chin 2001). The computer monitor segment of CRT production has historically grown in step with computer demand, but is expected to fall in coming years due to the increased popularity of liquid crystal displays. While the United States, Europe, and Japan were major producers of CRTs in the 1970s and 1980s, around 80 percent are now produced in East Asia (excluding Japan), especially in China and Taiwan.

Although CRTs have been in homes for decades and are probably taken for granted, CRT production processes are actually quite complex. The main component of a CRT is an evacuated glass tube with an electron gun mounted in the narrow end and a flat wide area opposite the gun that forms the screen. The glass in a CRT monitor is a specialty glass containing on average 8 to 9 percent lead in order to protect the user from electromagnetic radiation and improve optical qualities. The gun fires electrons that are then deflected by magnets to selectively hit different parts of the screen, which has phosphorescent coatings that glow different colors when struck. A mask (or matrix) of holes just behind the screen allows proper control of the imaging process.

Manufacturing the main glass tube is similar to other glassmaking, the main difference being the additional use of lead (EC 2001). The screen is much more complex as it must be coated with phosphors and a conductive material. As in chip fabrication, photolithography is used, this time to create a dense pattern of dots of different color phosphors. The pattern of holes or lines in the metallic mask behind the screen is also done with photolithography. The main tube and screen are joined, and then other components are added. The CRT is then placed in a steel frame, and chip-laden PCBs are installed to translate input from the computer or other video signal into the electron gun.

3.3.1.1 Materials and Energy Use in CRT Production

While recent information on materials use and emissions in CRT production is available from a study contracted by the U.S. Environmental Protection Agency (Socolof et al. 2002), here, data from the Japanese industry will be reported in order to make the material available in English. The results of a 1995 nation-wide survey of Japanese CRT manufacturers are shown in Table 3 (EIAJ 1997). The processes covered include the processing of masks, fluorescent coatings, and assembly and sealing of picture tubes, starting from specialized materials and parts. The use per CRT is estimated by dividing the Japanese total use by the total number of units produced (43.7 million in 1995). At the level of total materials use, picture tube production is not as intensive as microchip or circuit boards.

3.3.1.2 Environmental Impacts of CRT Production

The specific environmental issue usually raised for CRTs is the potential impact associated with the use of lead in CRT glass. One avenue of possible risk is lead exposure to workers involved in production of lead and CRT glass. This exposure naturally depends on the manufacturer's practices, and there are no readily available statistics or anecdotal evidence informing this issue. The other major potential avenue for lead impact is contamination of groundwater due to leaching of lead from CRTs in landfills. CRTs account for a significant 36 percent share of lead in landfills (IEC 1991), a fraction that is likely much higher today due to the addition of CRT monitors to the mix. CRT monitors fail laboratory leaching tests worse than circuit boards. Using the toxicity characteristic leaching potential (TCLP) test, one study indicated average leachable lead at 18.5 milligrams per liter (mg/l) from a sample of 36 CRTs (Townsend et al. 1999), while another obtained average results for four sets of 227 mg/l (Yang 1993). Both substantially exceed the U.S. regulated limit of 5.0 mg/l. The study by Yang also tested for leaching of other heavy metals and found that CRTs exceed regulated leaching limits for zinc as well (zinc is present due to its use in color phosphorescents). Based on such results, CRTs are classified as hazardous waste and subject to special handling requirements.

3.4 Liquid Crystal Displays (LCD)

3.4.1 The Global LCD Industry and its Production Processes

Liquid crystal display monitors continue to increase in popularity; some industry analysts believe shipments of LCD monitors may surpass their CRT counterparts by 2007. The significantly smaller size of LCD monitors is one reason consumers tend to prefer them. LCD production technology continues to improve rapidly, reducing the price and increasing the size and performance of screens. The global market for LCD monitors was valued at $9 billion and 12 million units in 2001. While Japan overwhelmingly dominated world production of LCD monitors in the 1990s, South Korea edged ahead in 2001 with a 40 percent market share, following by Japan (39 percent) and a rapidly expanding Taiwanese industry (21 percent) (Gerardino 2001).

The operating principle of an LCD is based on the phenomenon that certain special molecules, called twisted-nematic liquid crystals, can "twist" light depending on the electric field applied. Combined with polarized filters, this effect can be used to selectively block or pass light according to an applied electric signal. Image formation is achieved by sandwiching liquid crystal between two glass plates with transparent electrodes, on one side of which there is an array of small transistors, each of which controls a capacitor crossing the small column of liquid crystal. Light is shown through one side of the glass and the voltages applied to the transistors control how much passes through the liquid crystal column, creating a grayscale image. A color image is achieved by overlaying an array of red, green and blue filters.

Table 3: Materials and energy use to produce CRT picture tubes for TVs and monitors.

Category	Description	1995 Japan use		Use per CRT	
Raw materials	Processed glass (containing average 8% lead)	308,787	tons	7.1	kg
	Processed metal for panel	99,418	tons	2.3	kg
	Processed metal for yoke	15,443	tons	0.36	kg
	Processed metal for electron gun	2,098	tons	0.048	kg
	Processed ferrite	2,344	tons	0.054	kg
	Lead (for solder)	2,618	tons	0.060	kg
	Fluorescent compounds	634	tons	0.015	kg
	Plastics	1,467	tons	0.034	kg
	Total raw materials	432,809	tons	10	kg
Chemicals	Ferric chloride	7,906	tons	0.18	kg
	Hydrochloric acid	4,306	tons	0.10	kg
	Chlorine gas	2,359	tons	0.054	kg
	Sodium hydroxide	2,326	tons	0.054	kg
	Sulfuric acid	1,416	tons	0.033	kg
	Fluoric acid	577	tons	0.013	kg
	Hydrogen peroxide	575	tons	0.013	kg
	Other	1,933	tons	0.044	kg
	Total chemicals	21,398	tons	0.49	kg
Energy	Electricity	914.2 million	kWh	21	kWh
	Fossil fuels	1.33 million	liters oil	3	liters oil
	Embodied fossil	0.414 million	tons	9.5	kg
Water	Water	19.4 billion	liters	450	liters

Source: EIAJ (1997).

The starting point for LCD manufacturing is preparation of two thin glass plates. The strict requirements for flatness result in the use of specialized glass, usually soda lime or borosilicate glass. The electrode for the front plate is realized by sputtering, or shooting, molecules through a specialized "gun" of indium tin oxide. The patterned color filter is deposited by photolithographic processes similar to those described above in the section on semiconductors. The real challenge of the manufacturing process is deposition of the pattern of control transistors on the back plate. Each pixel on a screen requires three transistors, thus for a 1024 by 738 resolution screen, an array of 2.24 million transistors must be laid down over the area of the glass plate with no deviation or failures. This is technically challenging, and much of the high cost of LCD screens in the past was due to high failure rates. Manufacturing is similar to other semiconductor devices previously described, except that instead of a silicon wafer, the underlying base is a layer of doped amorphous silicon deposited via chemical vapor deposition. Rear lighting is achieved through four to eight miniature fluorescent bulbs.

3.4.2 Materials and Energy Use in LCD Production

The use and emissions of materials for LCD monitor production is described in a recent life cycle assessment commissioned by the U.S. Environmental Protection Agency. Process data was collected from surveys of seven Japanese manufacturers; details of use and emissions are reported in the study (Socolof et al. 2002). Table 4 abbreviates the results, displaying only aggregate use of chemicals and utilities for the panel manufacture and assembly of one LCD monitor. While wide fluctuations in energy data submitted by manufacturers were reported by the study authors, taking the results at face value suggests that the production of one LCD monitor requires 226 kilograms of embodied fossil fuels. This implies that manufacture is energy-intensive, much more so than for CRT monitors (see Table 3). Fabrication of the pixel-controlling array of transistors probably accounts for much of this energy use. One may expect, however, that rapid technological change in the industry, especially with regard to improved process yields, could bring this figure down considerably in the next few years.

Table 4: Aggregate chemicals, energy, and water use in the manufacture and assembly of a LCD monitor.

Material/input	**Amount used per monitor**
Photolithographic and other chemicals	3.7 kg
Elemental gases (N_2, O_2, argon)	5.9 kg
Electricity	87 kWh
Direct fossil fuels (98% natural gas)	198 kg
Embodied fossil fuels	226 kg
Water	1,290 liters

Source: Socolof et al. (2002).

3.4.3 Environmental Impacts in LCD Production

Manufacturing LCD screens utilizes many of the same processes as other semiconductor devices, thus similar environmental concerns arise. Therefore the main potential impacts on humans are long-term health effects due to chemical exposure. As explained earlier, these issues have yet to be resolved for the "traditional" semiconductor industry, and thus it is also unclear for the newer LCD technology.

Another possible environmental concern relates to the use of mercury in the fluorescent backlights of LCDs. Each LCD monitor contains about 4 to 12 milligrams of mercury; by comparison, a conventional fluorescent lamp for room lighting contains around 20 milligrams. The European Union is sufficiently concerned to include a criterion on mercury content for its eco-label for personal computers (Poll 2001).[9] To characterize the plausible scale of impacts of mercury, note that the total amount of mercury entering U.S. landfills in 2000 was estimated at 172 metric tons. An addition of one million LCD monitors would contribute 4 to 12 metric tons to this total, thus LCDs could become a more significant share of this total. Leaching from landfills, however, apparently represents a negligible fraction (less than 0.1 percent) of total mercury emissions in the United States; the figure is dominated by combustion of fossil fuels (87 percent) and co-production of chlorine and sodium hydroxide (10 percent) (EPA 1997). Thus the potential for LCDs to make an appreciable impact is apparently rather small.

The liquid crystals themselves are a unique constituent of LCD monitors. Many different compounds are used to form a mixture; generally, they are polycyclic aromatic hydrocarbons or halogenated aromatic hydrocarbons. What effects on human health might exposure to these chemicals pose? A study by manufacturers of liquid crystals reported that out of 588 compounds used, only 26 posed acute toxic potential (Socolof et al. 2002). There have apparently been no tests yet of the carcinogenic potential of these compounds in animals; tests on bacteria indicated no suspicion of mutagenic effects (ibid.). The study has yet to be made publicly available, however, and it is reasonable to assert that such work should be submitted for peer-review in order to gain public acceptance.

3.5 Bulk Materials in a Computer

There are also environmental impacts associated with producing the bulk materials contained in a computer. The various processes and environmental impacts associated with producing these will not be surveyed here; traditional industries such as steel and glass are treated elsewhere in a variety of contexts.

The treatment of bulk materials here is a description of their contribution to the material composition of a PC and estimation of energy use in producing one computer. The data and results of estimation for the "control unit" (i.e., the tower) of a desktop computer and CRT monitor appear in Table 5 and 6. The energy intensities of materials were taken from standard databases used for life cycle assessment (Boustead 1999; BUWAL 1997; Design for Sustainability Program 2001), except for silver and gold, which had no listed values. The energy intensity for producing gold and silver was taken from industry data on energy use at the Australian Olympic Dam mine, dividing energy use for the three co-product metals (uranium, gold, and silver) according to their economic value (WMC 1998). Data on the materials content of a

[9] See Chapter 8 for definition and detailed discussion of eco-labels.

control unit came from Shimoda and collaborators (which refer to an NEC database) and a report from Miyamoto and collaborators on a CRT monitor (Shimoda et al. 1998; Miyamoto et al. 1998). Although the content of gold and silver in a PC is rather small, they were included because their mining has a high environmental burden per kilogram, and thus should be checked even if the amounts used are small. Presuming the data used are reasonable, the result suggests that the amount of energy consumed to produce the gold and silver used in PCs is not significant.

Plastics are listed as an aggregate category, and in practice a mixture of different varieties of plastics are used. A sample composition of plastics in a desktop control unit by weight is polystyrene (PS), 33 percent; polyphenylene ether (PE), 20 percent; acrlonitril-butadiene-acrylate (ABS), 19 percent; polyvinyl chloride (PVC), 15 percent; polybutylene terephthalate (PBT), 9 percent; polyacetal (POM), 3.5 percent; and polyethylene terephthalate (PET), 0.5 percent (Shimoda et al. 1998). The variety of plastics used continues to pose a challenge to their recovery in recycling processes.

Note that the energy figures in Table 5 and 6 refer only to the amount needed to produce the "standard" industrial grade material and do not include further processing to specialized forms. Purification and processing energy costs can be substantial, and will be partially addressed in the next section.

The total energy used to produce the materials in a desktop control unit and CRT monitor, 770 megajoules and 800 megajoules, corresponds to burning 21 and 22 kilograms of fossil fuels, respectively.[10] Since these figures are much smaller than those for production of chips, circuit boards, and process chemicals, the results suggest that the high-tech components in computers are far more environmentally relevant than the bulk materials.

3.6 Specialized Chemicals and Materials for Electronics Manufacturing

3.6.1 The Global Industry for Specialized Chemicals and Materials

A variety of specialized chemicals and materials are used in electronics manufacturing, such as silicon wafers, photoresists, etchants, dopants and other chemicals for making microchips and circuit boards. The materials and chemicals are generally extra high-grade. Chemicals for semiconductor fabrication typically run in the range of 99.999-99.9995% pure, compared to industrial grades which run in the 90-99% range.

[10] This assumes that 37 MJ equals 1 kg fossil fuel, which is obtained by dividing the global supply of energy delivered (including electricity) by the weight of worldwide consumption of fossil fuels in 2000 (IEA 2002).

Table 5: Content and associated energy use in production of materials for desktop control unit.

Material	Energy intensity of material (MJ/kg)	Main use(s) in control unit	Amount contained (grams)	Energy content (MJ/unit)
Steel	59	Housing	6,050	360
Copper	94	Wires, circuit boards	670	63
Aluminum	214	HD, circuit boards	440	94
Plastics	84	Housing, CD-ROM	650	55
Epoxy	140	Circuit boards	1,040	150
Tin	230	Solder	47	11
Lead	54	Solder	27	1.5
Nickel	340	Disk drive	18	6.2
Silver	1570	Circuit boards	1.4	2.3
Gold	84,000	Circuit boards	0.36	30
Subtotal			8,944	770
Other			96	
Total			9,040	770

Sources: Shimoda et al. (1998), Boustead (1999), BUWAL (1997), Design for Sustainability Program (2001), and WMC (1998).

The total global market in 1999 for chemicals and materials used in the production of semiconductors and circuit boards totaled $22 billion, the value of different classes of chemicals and materials are detailed in Table 7. From 1999, the wafer industry was forecast to grow at 11.4 percent per year and chemical sectors around 8 percent per year (Electronic Chemical News 2000). These growth rates are slower than those of the destination sectors—average growth of the semiconductor and circuit board markets are 16 percent and 11 percent per year, respectively. The difference in growth rates is presumably due to technological progress and greater efficiency in the industry, which reduces the amount of material needed per unit of production.

3.6.2 Energy use in producing specialized chemicals and materials

Producing extra-high grade materials entails additional environmental impacts. All chemical inputs to semiconductor processes must thus go through rigorous purification processes, often based on vacuum distillation, which is well known to be energy intensive. The variety and specialized nature of chemicals and materials used in electronics manufacturing makes characterization of the required materials and energy challenging. Publicly available process data for production of such materials is almost non-existent; all standard databases cover only the "standard" industrial grade of chemicals and materials. Energy use in the production of silicon wafers was studied in previous work by the author and collaborators; the results were that the chain of processes involved in the manufacture of a wafer consume 2,150 kilowatt-hours per kilogram (0.34 kWh/cm^2) of contained silicon—160 times more energy than that required to produce "standard" grade silicon (Williams et al. 2002).

Given existing work describing energy used in silicon wafer production processes, the energy investment to produce the wafers contained in a computer can be estimated according to the following equation (using 1995 data):

Equation 1

Silicon wafer energy / computer [kWh/unit] =
(Energy to produce wafers X world wafer production X computer share) / computers produced =
(0.34 kWh/cm^2 X 28.4 billion cm^2 X 49%) / 90 million units =
53 kWh per computer.

Applying the conversion factor of 320 grams of fossil fuel per kilowatt-hour of electricity (as was used to estimate embodied fossil fuel use in Tables 1 to 6), this corresponds to an embodied fossil fuel use of 17 kilograms to produce the silicon wafers contained in a computer.

Table 6: Content and associated energy use in production of materials for 17 inch CRT monitor.

Material	Energy intensity of material (MJ/kg)	Main use(s) in CRT	Amount contained (grams)	Energy content (MJ/unit)
Glass	15	Picture tube	6,817	100
Steel	59	Housing	2,830	170
Copper	94	Wires, circuit boards	700	66
Ferrite	59	Deflection yoke	480	28
Aluminum	214	Heat sinks	240	51
Plastics	84	Housing	3,530	300
Epoxy resin	140	Circuit boards	140	20
Tin	230	Solder (circuit boards)	20.0	4.6
Lead	54	Glass, solder	593	32
Silver	1,570	Circuit boards	1.24	1.9
Gold	84,000	Circuit boards	0.31	26
Subtotal			15,352	800
Other			98	
Total			15,450	800

Sources: Miyamoto et al. (1998), Boustead (1999), BUWAL (1997), Design for Sustainability Program (2001), and WMC (1998).

What about chemicals and materials for electronics manufacturing other than silicon wafers? Given the unavailability of data on production processes for these materials, only a crude calculation of energy use can be attempted here. Instead of using facility-level data, it is also possible to estimate energy use (and emissions) to make products using macro-economic input-output tables. The most detailed national input-output tables divide an economy into 400 to 500 distinct sectors and describe monetary transactions between these sectors. Methods that combine this information with macro data on sectoral energy use have been developed to estimate the total energy per dollar consumed by the supply chain for a given sector (Bullard and Herendeen 1975). The use of prices allows such results to be applied for products as well as entire industrial sectors; in this context the technique is termed economic input-output lifecycle assessment (Hendrickson et al. 1998).

Table 7: Markets for specialized chemicals and materials used in electronics manufacturing.

Process	**Item**	**Global market (million U.S.$)**
Semiconductor fabrication	Silicon wafers	5,600
	Photoresists, developers, ancillaries	1,632
	Solvents, acids	728
	Metals for deposition	377
	Gases	1,410
	Photomasks	830
	Other	955
	Subtotal	11,532
Semiconductor packaging	Lead frames	2,330
	Encapsulents (epoxy)	869
	Ceramic packages	1,545
	Micro-thin bonding wires	444
	Other	507
	Subtotal	5,695
Wiring board fabrication	Photolithographic chemicals	1,168
	Solder masks	564
	Copper deposition chemicals (electrolytic and electroless)	481
	Other	205
	Subtotal	2,418
Wiring board assembly	Solder and solder fluxes	2,193
	Cleaning agents	489
	Other	131
	Subtotal	2,813
Total semiconductor and PWB materials		22,458

Sources: Electronic Chemical News (2000) and Van Arnum (2000).

The largest uncertainty in this estimate relates to choosing a macro-economic sector that most closely matches the products under consideration, in this case chemicals and materials for electronics manufacturing. None of the 480 sectors in the 1997 U.S. input-output table are an obvious match. The three closest possibilities and their supply chain energy requirements are listed in Table 8. The sector "organic and inorganic chemicals" might seem appropriate at first glance, but the main part of production is bulk chemicals, which require considerable energy for small value added. Instead, the "chemicals and chemical preparations" sector is chosen to represent production of electronic chemicals and materials. This is probably an overestimate; as a safeguard the sector "electrical machinery and supplies" is used to suggest a plausible lower bound.

Table 8: Supply chain energy requirements to make U.S.$1 value of products in three sectors.

Case	Sector from 1997 U.S. 480x480 input-output table	Direct fossil fuel use (MJ/$)	Electricity (kWh/$)	Embodied fossil fuels (kg/$)
Base	Chemicals and chemical preparations, n.e.c.*	21.65	1.07	0.88
Lower bound	Electrical machinery, equipment, and supplies, n.e.c.	5.60	0.52	0.31
Upper bound	Organic and inorganic chemicals	33.43	1.15	1.21

Source: Green Design Initiative (2003).
*Note: n.e.c. stands for "not elsewhere classified."

Given this discussion, the estimate of energy needed to produce electronic materials for a computer is done using the following formula (using 1999 data):

Equation 2

Fossil fuels to produce electronic materials per computer [kg/unit] =
(global market for electronic materials [$] X energy intensity of sector [kg/$] X fraction of materials used for computers [%])/ computers produced [units]
= $17 billion X .88 kg/$ X 49% / 114 million units = 64 kg fossil fuels

$17 billion is the total market for electronics chemicals and materials excepting silicon wafers, and the fraction of total energy used in producing chemicals and materials for electronics allotted for computers only is assumed to be 49 percent—identical to the share of the worldwide semiconductor market destined for computers. The global production of computers in 1999 was 114 million units. Combining these data and assumptions, the embodied fossil fuels for electronic materials used to produce a computer is estimated at 64 kilograms for the base case, with a range of 22 to 87 kilograms from lower and upper bounds.

3.6.3 Environmental Impacts of Producing Chemicals and Materials for Electronics Manufacturing

Hundreds, even thousands, of chemicals and other materials are used to manufacture electronic products. It is therefore not feasible to discuss all of their environmental impacts in great detail here. The main point is that the consideration of the risks associated with manufacturing semiconductors and circuit boards must also include the processes to make the specialized chemical and material inputs to make these components. These sectors are almost always neglected in existing analyses, which typically only consider crude raw materials (such as steel) and final production of components. There are some practical factors that explain why intermediate materials and processing tend to "fall through the cracks"—the sheer diversity of types and grades of chemicals and materials is daunting, to say the least. It is also fair to argue, however, that little attention has yet to be given to the challenge of gathering data and understanding the environmental impacts of these industries supplying materials to the electronics industries.

4. TOTAL FOSSIL FUELS, CHEMICALS, AND WATER USED IN THE PRODUCTION OF A DESKTOP COMPUTER

The various components of the computer production chain can now be combined here to give a rough picture of total amounts of materials and energy associated with the manufacture of a computer. Combining the previous section's findings, 240 kg of fossil fuels, 22 kg of chemicals, and 1,500 kg of water are needed to make one desktop computer with a 17 inch CRT monitor. The details of the calculation are shown in Table 9. Note that 240 kilograms of fossil fuels corresponds to 5,040 megajoules in standard energy units.

The amount of energy used to produce a computer is significant when compared with other consumer goods. An automobile or refrigerator requires around 2,000 kilograms and 50 kilograms of fossil fuels to produce, respectively (MacLean and Lave 1998; EC 1999). The short life span of the typical computer makes this high amount of energy in production even more noteworthy. Consumers often purchase a new computer every two years (see Chapter 9, section 3), compared to every ten years for an automobile or refrigerator; thus five times as many computers are needed over a ten-year period. The production of a computer also requires more energy to produce relative to its weight than an automobile or refrigerator. The ratio of embodied fossil fuels to product weight is nine times for a computer, but only one to two times for an automobile or refrigerator. The reason that computers use significantly more energy per weight is because of the high concentration of high-tech parts such as semiconductors; as well, the complex internal structure of microchips requires extensive processing.

The fossil fuel consumption over the entire life cycle of production and use of a computer (1,470 kg) is close to that of a refrigerator (1,330 kg).[11] The share of energy use during the phases of the life cycle is entirely different, however; for a refrigerator, 96 percent is consumed during the use phase, while for the computer this figure is 25 percent. It follows that increasing energy efficiency during use is a natural strategy to follow for reducing the life cycle energy use of refrigerators; while for computers, reducing energy use in production and extension of useable life span stand out as especially effective options.

---

[11] This calculation assumes energy consumption of a refrigerator to be 400 kWh/year and a home computer, 117 kWh/year.

4.1 Review of Other Studies of Total Resource Use for Producing Computers

It is useful to compare these figures with other computer studies reported in the literature. There is a trend towards "black box" studies, that is, announcements of only final results without substantive information on data and assumptions used. For a study to be credible, it requires a thorough description of the computer being assessed, the processes included, and breakdowns of energy use for major components of the computer. While no existing (English language) study fully meets these criteria, two well-known ones are worth mentioning. The Microelectronics and Computer Technology Corporation (MCC 1993) published results of a landmark life cycle study of a computer workstation in 1993; the team found that 2,300 kilowatt-hours of electricity are needed to produce one computer. This translates into 740 kilograms of fossil fuels per computer. The result is significantly higher than obtained here, perhaps because technological progress made during the intervening years (the MCC data is probably from the late 1980s). The European Union commissioned a 1998 study that showed that 3,630 MJ (92 kg of fossil fuels) and 2.6 million kg of water are needed to produce a desktop computer with CRT monitor (Atlantic 1998). The latter figure for water use is an obvious overestimate as it would imply world computer production in 2000 of 120 million computers requires about 40% of world industrial water use. A possible explanation of why the energy figure is lower than the results here is probably because the Atlantic Consulting/IPU analysis does not include the energy used to produce silicon wafers and electronic chemicals and materials. It is difficult to further examine why the MCC and E.U. studies give different results than those here, as neither report contains quantitative information on process data used.

Table 9: Fossil fuels, chemicals, and water consumed in the production of one desktop computer.

Item	Fossil fuels (kg)	Chemicals (kg)	Water (kg)
Semiconductors	94	7.1	310
Printed circuit boards	14	14	780
CRT picture tube	9.5	0.49	450
Bulk materials - control unit	21	NI	NI
Bulk materials - CRT	22	NI	NI
Electronic materials/chemicals (excluding wafers)	64	NI	NI
Silicon wafers	17	NI	NI
Manufacture of parts	NI	NI	NI
Assembly of computer	NI	NI	NI
Total	240	22	1,500

Notes: NI = not included in analysis. Only two significant digits have been kept in sums.

Another approach to estimate the use of embodied energy/fossil fuels uses economic input-output analysis. The supply chain energy intensity of the U.S.

electronic computer sector from 1997 tables is 0.31 kilowatt-hours of electricity use per dollar of computers produced and 4.8 megajoules for fossil fuels per dollar (Green Design Initiative 2003). Using $1,100 as the average 1998 price for a personal computer, the estimate for embodied fossil fuels to produce one computer amounts to 240 kilograms (or 5,600 MJ).[12] This result is surprisingly close to the estimate obtained here. Why the two very different methods give close results is too technical and lengthy a topic for this chapter, suffice it to say that an overall conclusion that can be drawn from this review of studies is that the figure for fossil fuels used to produce one desktop computer in the late-1990s was probably in the 200- to 300-kilogram range.

5. CONCLUSIONS

High energy use, possible health effects due to exposure to particular substances used in computer production, and hazardous substance content are three important environmental issues associated with PCs. Many questions remain unresolved, such as the effects on cancer rates of working in semiconductor fabrication facilities.

Progress is being made towards understanding the effects of brominated flame retardants; a number of research groups around the world are committed to continuing studies on epidemiology and exposure routes. The research picture is less promising regarding health effects on workers in the semiconductor industry. No significant (public) studies are currently underway on that topic. Such research would require active cooperation from the industry, which is not likely to be forthcoming without pressure from governments or the public.

The lack of consensus on the health effects of heavy metals used in computers is not due to a lack of basic scientific knowledge. The health impacts of lead and mercury and the chemistry of leaching from landfills are hardly new fields of study, but existing knowledge has yet to be applied to assess the risk associated with heavy metals used in electronics. This situation is surprising, given that legislation in the European Union will ban the use of these substances in electronic equipment starting in 2006 and many companies are already touting their products as eco-friendly because they avoid use of these substances. No major moves to shed more light on this issue are evident, although a U.S. EPA-sponsored project that will implement a life cycle assessment comparing the environmental impacts of lead and lead-free solders (EPA 2002). Apparently, that study will not include a macro-level risk assessment of lead solders, however. Undertaking risk assessments should automatically be made part of policy formulation processes—an issue discussed further in Chapter 13.

Even with today's limited understanding of the key issues, many actions can still be taken to reduce the environmental impacts of computers. One important action is to extend the length of time we use information technology (IT) equipment—the short lifespan drives additional production and the generation of waste such that all

---

[12] $1,100 × (31 kWh/$ × 320 g/kWh + 4.8 MJ/$ ÷ 40MJ/kg) = 240 kg (to two significant digits).

associated environmental impacts are magnified. While technological change clearly drives the rapid cycle of production and disposal, a great deal of IT equipment is certainly "wasted"—many devices are thrown away that are still functional and in demand by certain groups of users. Public awareness and response, government action to facilitate, and a pro-active attitude from firms can do much to increase the utilization of IT equipment. These issues are discussed further in other chapters in this book. Also, manufacturers can further reduce the energy needed to produce computers by improving systems to measure and reduce energy use in the supply chain. Many opportunities exist to reduce the environmental burdens of computers for little or no economic cost, it is a matter of gathering the will to take action.

5.1 Non-technical Summary

The manufacture of computers entails a variety of environmental impacts associated with processes in the production chain. Three main areas of impacts are:

- Significant energy use in the chain of production processes to make computers.
- Possible long-term health effects on workers, families, and neighboring communities due to chemical exposure and emissions from production stages such as microchip fabrication.
- Possible health impacts due to exposure to hazardous materials contained in computer products, in particular brominated flame retardants, lead, and mercury. The main risk of exposure is probably from computers that have been disposed in landfills at their end of life.

The production of IT equipment is energy-intensive—thus significant for climate change and the depletion of fossil fuel resources. This study estimated the total fossil fuels used to manufacture one desktop computer at 240 kilograms (530 lb). This suggests that computers are much more energy-intensive than many other products we use in daily life. The fossil fuels used to produce a computer are about nine times its weight. By comparison, the amount of fossil fuels needed to produce an automobile or refrigerator is around twice their weights. The production of a computer is more energy-intensive due to its high-tech nature; extensive processing is required to create its complex components. The combined high-energy intensity and relatively short useable life span of a computer raise its lifetime environment-related energy impacts to about the same level as a refrigerator—one of the more energy-intensive appliances in the home. But whereas nearly all of a refrigerator's lifetime fossil fuel consumption (96 percent) is due to the electricity produced to run it and only a small amount for production, for a computer the situation is reversed; a relatively small proportion (25 percent) is for use, while most (75 percent) is for production. In contrast with other appliances such as refrigerators, where improving electricity efficiency during operation is the main strategy to reducing energy use over the life cycle, for computers it is very important to save energy in the production phase and increase the life span of the use phase.

Hundreds or even thousands of chemicals, many toxic, are used to produce a computer. This study estimated the total weights of chemicals and water needed to produce one desktop computer to be 22 kilograms (48 lb) and 1,500 kilograms (3,300 lb), respectively. The high water usage reflects the tendency of high-tech processes to require frequent rinsing during chemical processes. A set of specific health concerns has arisen regarding chemical exposure in to the production of computers. There are indications that working in microchip production plants may elevate rates of cancer and birth defects. This issue is still being debated in the context of a number of lawsuits in which former workers are suing semiconductor firms. Scientific evidence cannot yet prove or disprove the claims of workers; no large-scale study of links between working in a semiconductor factory and cancer rates, for example, has been undertaken. Another chemical-related concern is that emissions (especially accidental leakages) of toxic chemicals from factories can affect nearby residents and ecosystems.

Policymakers have given much attention in recent years to addressing the environmental impacts of emissions of hazardous substances from disposed IT equipment. A group of chemicals called brominated flame retardants is of particular concern. These chemicals, including polybrominated diphenyl ether (PBDE), are used in circuit boards and plastic cases for computers to make them less flammable. Recent studies have shown that concentrations of these compounds in humans have been increasing rapidly in Sweden and Canada and the United States. In sufficient dosage these chemicals cause neuro-developmental disorders and possibly cancer. It is not yet clear, however, whether current or projected levels of exposure to these agents are significant enough to cause health effects. The pathways of exposure to these chemicals are also poorly understood and, as other common products such as furniture often contain the same flame retardants, the relative impacts of their use in computers is not yet known.

Computers and monitors also contain heavy metals such as lead, mercury, cadmium, and chromium. These substances represent potential health risks to workers during computer production and environmental risks to water supplies near landfills where used electronic products have been dumped. The lead content of cathode ray tube (CRT) displays is significant (hundreds of grams per monitor), and makes up around 36 percent of lead that finds its way into municipal landfills in the United States. The use of lead-containing solder to attach components to circuit boards has become a contentious issue now that the European Union has passed legislation that will ban the use of lead in solder starting in 2006. Some argue that there has been no real attempt to quantify the environmental impacts of lead in circuit boards. Existing studies do not show the leaching of lead from landfills to be a particular problem, and circuit boards represent only around 4 percent of lead inputs to landfills. Mercury, cadmium, and chromium are present in computer equipment in much smaller amounts, mainly in the small fluorescent bulbs of liquid crystal display (LCD) monitors.

Societal response is needed both to clarify the environmental impacts of computers and also to reduce them. With respect to understanding the problems, scientific work continues in some areas, such as research to clarify the health effects of brominated flame retardants. But many issues, such as health impacts on workers involved in semiconductor fabrication, are still at a stalemate. Demands from governments and civil society may build pressure to conduct the needed studies. Although the potential health effects of heavy metals leaching from landfills can already be estimated using existing scientific knowledge, policymakers continue to develop policies and governments implement them in the absence of proper risk assessments.

Despite our incomplete current understanding of the health- and environment-related impacts of computers, today's computers users can already take many actions to improve the situation. Most important are actions that improve the utilization of existing equipment—that is, to extent its useful life. Users should consider carefully whether it is really necessary to purchase a new computer, upgrade the existing computer instead to meet current needs when possible, and promptly sell the old computer to the used-product market when they do decide to dispose of it. In these ways, computer users can make a big difference by reducing the number of computers produced and eventually ending up in landfills—and thereby reduce their impacts on the environment.

REFERENCES

Atlantic Consulting/IPU. 1998. *LCA study of the product group personal computers in the EU Ecolabel scheme*, version 1.2.1. London: Atlantic Consulting and IPU.

Ayres, R. and L. Ayres. 2002. *Handbook of industrial ecology*. Cheltenham, UK: Edward Elgar.

Bailar, J., M. Bobak and M. Fowler. 2000. Open letter to the Health and Safety Executive. *Int. J. Occup. Environ. Health* 6:71–72

Bailar, J., M. Greenberg, R. Harrison, J. LaDou, E. Richter and A.Watterson. 2002. Cancer risk in the semiconductor industry: A call for action. *International Journal of Occupational & Environmental Health* 8(2): 163–168.

Bernama. 1998. Japan: Toshiba reports toxic leakage near 3 Kawasaki plants. *Bernama*, June 5.

Boustead. 1999. Boustead Model for life cycle inventory calculations, version 4.2. West Sussex, UK: Boustead Consulting. <http://www.boustead-consulting.co.uk/products.htm> (10 January 2003).

Bullard, C. and R. Herendeen. 1975. The energy cost of goods and services. *Energy Policy* (December): 268–277.

BUWAL (Bundesamt für Umwelt, Wald und Landschaft. 1997). *Economic Inventory of Packaging Database* (also known as BUWAL 250). Zurich: Swiss Federal Office of Environment, Forests and Landscape and the Swiss Packaging Institute.

Chin, S. 2001. LCDs creeping up on CRTs as prices fall. *EBN* 1288 (12 November): 50.

Circuits Assembly. 2001. PWB and flex circuit production up. *Circuits Assembly* 12(9): 16.

Correra A., R. Gray and R. Cohen. 1996. Ethylene glycol ethers and risks of spontaneous abortion and subfertility. *Am. J. Epidemiol.* 143:707–717.

Crothers, B. 1999. Surveys: World PC shipments rebound. CNETnews.com. <http://news.com.com/2100-1001-220898.html> (20 January 2003).

Curran, M. 1996. *Environmental life-cycle assessment*. New York: McGraw-Hill.

Dapson S. C., D. E. Hutcheon and D. Lehr. 1984. Effect of methyl chloroform on cardiovascular development in rats. *Teratology* 29(2): 25A.

Darnerud P., S. Atuma, M. Aune, S. Cnattingius, M. Wernroth and A. Wicklund-Glynn. 1998. Polybrominated diphenyl ethers (PBDEs) in breast milk from primiparous women in Uppsala county, Sweden. *Organohalogen Compd.* 35:411–414.

Darnerud, P., G. Eriksen, T. Johannesson, P. Larsen and M. Viluksela. 2001. Polybrominated diphenyl ethers: Occurrence, dietary exposure, and toxicology. *Environ. Health Perspect.* 109 (suppl 1): 49–68.

Design for Sustainability Program. 2001. IDEMAT Database. Delft University of Technology, Faculty of Design, Engineering and Production. <http://www.io.tudelft.nl/research/dfs/idemat/> (21 January 2003).

Electronic Chemical News. 2000. Semiconductor chemicals to grow 13%/year. *Electronic Chemical News* 15(17): 5.

Electronics Industry Association of Japan (EIAJ). 1997. *Denshibuhinsangyou kankyou bijon* (Environmental vision for the electronics parts industry). Tokyo: Electronics Industry Association of Japan.

Environmental Protection Agency (EPA). 1995. *Printed wiring board industry and use cluster profile.* Design for environment program. Washington, D.C.: U.S. Environmental Protection Agency. <http://www.epa.gov/opptintr/dfe/pubs/pwb/tech_rep/usecluster/index.html> (14 January 2003).

———. 1997. *Mercury study report to Congress,* vol. 1, executive summary. EPA-452/R-97-003. Washington, D.C.: U.S. Environmental Protection Agency. <http://www.epa.gov/airprogm/oar/mercury.html> (20 January 2003).

———. 2002. Assessing life-cycle impacts, lead-free solder partnership. EPA 744-F-02-007. Washington D.C.: United States Environmental Protection Agency. <http://eerc.ra.utk.edu/ccpct/lfsp-docs/solder-factsheet3.pdf> (21 December 2003).

European Commission. 1999. *Revision European Eco-label criteria for refrigerators,* draft final report. EC DG XI.E.4. <http://www.vhk.nl/download/Ecolabel_Fridge_draft_report.pdf> (20 January 2003).

———. 2001. Reference document on best available techniques in the glass manufacturing industry. Integrated Pollution Prevention and Control (IPPC), European Commission. <http://www.sepa.org.uk/guidance/ippc/bref/PDFs/glass.pdf> (15 January 2003).

European Parliament. 2003. Directive 2002/95/ED of the European Parliament and of the council of 27 January 2003 on the restriction of the use of certain hazardous substances in electrical and electronic equipment (RoHS). *Official Journal of the European Union* L 37:19–23.

Fisher, J. 2001. Poison valley: Is workers' health the price we pay for high-tech progress? *Salon.com* <http://archive.salon.com/tech/feature/2001/07/30/almaden1/> (15 January 2003).

Gerardino, K. 2001. Japan sees falling LCD sales. *EE Times,* 28 September. <http://www.eetimes.com/story/OEG20010926S0061> (20 January 2003).

Green Design Initiative. 2003. *Economic input-output life cycle assessment.* Pittsburgh, PA: Carnegie Mellon University, Green Design Initiative. <http://www.eiolca.net/> (5 September 2002).

Griese, H., J. Muller, H. Reichl, G. Somi, A. Stevels and K. H. Zuber. 2000. Environmental assessment of lead free international systems. Proceedings of the Symposium on Lead Free Interconnect Technology, SMTA, 13–14 June 2000, at Boston, MA, USA.

Hellwege, J. 2002. Coming clean: Workers sue computer chip makers over chemical exposures. *Trial* (March): 12–17.

Hendrickson, C. T., A. Horvath, S. Joshi and L. B. Lave. 1998. Economic input-output models for environmental life-cycle assessment. *Environmental Science and Technology* 32(4): 184A–191A.

International Energy Agency (IEA). 2002. Key world energy statistics 2002. Paris: International Energy Agency. <http://www.iea.org/statist/keyworld2002/key2002/keystats.htm> (20 May 2003).

IEC. 1991. Potential human exposures from lead in municipal solid waste. Prepared for Lead Industries Association. Industrial Economics, Inc. New York: IEC.

Jakobsson, K., K. Thurreson, L. Rylander, A. Sjodin, L. Hagmar and A. Bergman. 2002. Exposure to polybrominated diphenyl ethers and tetrabromobisphenol among computer technicians. *Chemosphere* 46:709–716.

Kanegsberg, B. and P. Van Arnum. 1999. Solvents for wafer surface preparation. *Chemical Marketing Reporter* 256 (1) (5 July).

MacLean, H. and L. Lave 1998. A life-cycle model of an automobile. *Environmental Science & Technology* 32(13): 322A–329A.

Mainichi. 1998. Carcinogens found in ground water at 4 Osaka factories. *Mainichi Daily News*, 25 June.

Microelectronics and Computer Technology Corporation (MCC). 1993. *Environmental consciousness: A strategic competitiveness issue for the electronics and computer industry*. Austin, TX: Microelectronics and Computer Technology Corporation.

Miller, H. 2002. *Lead-free electronic solder: Why?* Palo Alto, CA: InfraFOCUS.

Miyamoto, S., M. Tekawa and A. Inaba. 1998. Life cycle assessment of personal computers for the purpose of design for environment (in Japanese). *Energy and Resources* 19(1): 75–80.

NUS Corporation. 1988. Summary of data on municipal solid waste landfill leachate characteristics. Prepared for the U.S. EPA, Washington, D.C.

Pastides, H., F. Calabrese, D. Hosmer and D. Harris. 1988. Spontaneous abortion and general illness symptoms among semiconductor manufacturers. *J. Occup. Med.* 30:543–551.

Poll, J. 2001. *Revision of the EU Ecolabel criteria for computers*. European Union, AEAT/ENV/R/0751. <http://europa.eu.int/comm/environment/ecolabel/pdf/personal_computers/finalreport_aug2001.pdf> (20 January 2003).

Ryan, J. and B. Patry. 2000. Determination of brominated diphenyl ethers (BDEs) and levels in Canadian human milk. *Organochlorine Compd.* 47:57–60.

Schenker, M., E. Gold and J. Beaumont et al. 1995. Association of spontaneous abortion and other reproductive effects with work in the semiconductor industry. *Am. J. Ind. Med.* 28:639–659.

Semiconductor Industry Association (SIA). 2002. *Industry statistics*. San Jose, CA: Semiconductor Industry Association. <http://www.sia-online.org/pre_statistics.cfm> (28 December 2002).

She, J., J. Winkler, P. Visita, M. McKinney and M. Patreus. 2000. Analysis of PBDEs in seal blubber and human breast adipose tissue samples. *Organohalogen Compd.* 47:53–56.

Shimoda, Y., N. Yoshida, N. Yasuda, T. Shirakawa and T. Morioka. 1998. LCA for personal computer in consideration of various use and upgrading. Proceedings of the Third International Conference on Ecobalance, 25–27 November 1998, at Tsukuba, Japan, 251–254. Tokyo: Society for Non-Traditional Technology.

Smith, E. and L. Kristine Swanger. 1999. Lead free solders—A push in the wrong direction? Paper presented at IPC Printed Circuits Expo, 14–18 March 1999, at Long Beach, CA.

Socolof, M., J. Overly, L. Kincaid and J. Greibig. 2002. *Desktop computer displays: A life-cycle assessment*. EPA 744-R-01-004. Washington D.C.: U.S. Environmental Protection Agency. <http://www.epa.gov/oppt/dfe/pubs/comp-dic/lca/index.htm> (15 January 2003).

Stuart, A. F. H. 1997. The economic significance of the semiconductor industry. In *Spectator or serious player? Competitiveness of Australia's information industries*, appendix F. Australian government DIST study, reporting Garnter/Dataquest data. <http://www.dist.gov.au/itt/tskforce/allen/index.html> (28 December 2002).

Swan, S. et al. 1985. Pregnancy outcomes in Santa Clara County, 1980–1982: Summaries of two epidemiological studies. Berkeley, CA: California Dept. of Health Services.

Townsend, T., S. Musson, Y. C. Jang and I. H. Chung. 1999. Characterization of lead leachability from cathode ray tubes using the toxicity characteristic leaching procedure. Report #99-5. Gainesville, FL: Florida Center for Solid and Hazardous Waste Management. <http://www.enveng.ufl.edu/homepp /townsend/Research/CRT/CRTDec99.pdf> (15 January 2003).

Townsend, T., Y. Jang, T. Tolaymat and J. Jambeck. 2001. Leaching tests for evaluating risk in solid waste management decision making. Draft report. Gainesville, FL: Florida Center for Solid and

Hazardous Waste Management. <http://www.enveng.ufl.edu/homepp/townsend/Research/Leach/Leach_Yr1.PDF> (8 January 2003).

Turbini, L., G. Munie, D. Bernier, J. Gamalski and D. Bergman. 2000. Assessing the environmental implications of lead-free soldering. Proceedings of Joint International Congress and Exhibition: Electronics Goes Green 2000+, 11–13 September 2000, at Berlin, 37–42. VDE Verlag.

U.S. Census Bureau. 2001. *Annual survey of manufactures, statistics for industry groups and industries (1993–2001)*. U.S. Department of Commerce, Bureau of the Census. <http://www.census.gov/mcd/asmhome.html> (28 December 2002).

Van Arnum, P. 2000. Electronic materials. *Chemical Market Reporter* (11 December): FR8–FR12.

Van Zant, P. 1997. *Microchip processing*. 3d ed. New York: McGraw-Hill.

Wenning, R. 2002. Uncertainties and data needs in risk assessment of three commercial polybrominated diphenyl ethers: Probabilistic exposure analysis and comparison with European Commission results. *Chemosphere* 46:779–796.

Williams, E. 2000. *Global production chains and sustainability: The case of high-purity silicon and its applications in information technology and renewable energy*. Tokyo: United Nations University Institute of Advanced Studies. <www.it-environment.org/publications.html> (28 December 2002).

Williams, E., R. Ayres and M. Heller. 2002. The 1.7 kg microchip: Energy and chemical use in the production of semiconductors. *Environmental Science & Technology* 36(24) (15 December): 5504–5510.

WMC. 1998. *Greenhouse challenge: Summary report 1998*. Southbank, Australia. WMC Resources Ltd. <http://www.wmc.com.au/sustain/environ/greenhouse/greenhouse.pdf> (10 January 2003).

Yamauchi, I., ed. 2000. *Shinkinzoku Deetabukku* (New metals databook). Tokyo: Homat Ad.

Yang, G. 1993. Environmental threats of discarded picture tubes and printed circuit boards. *Journal of Hazardous Materials* 34(2): 235–243.

Yoshida, F. 2002. *The economics and waste and pollution management in Japan*. Tokyo: Springer-Verlag.

Yu-Tzu, C. 2002. Official may resign over water crisis. *Taipei Times*, 7 May. <http://www.taipeitimes.com/News/front/archives/2002/05/07/134933> (8 January 2002).

Chapter 4

HOW THE EUROPEAN UNION'S WEEE DIRECTIVE WILL CHANGE THE MARKET FOR ELECTRONIC EQUIPMENT—TWO SCENARIOS

Klaus Hieronymi[a] and Axel Schneider[b]

[a]*Hewlett-Packard, Europe*
[b]*Promotionteam, Germany*

1. INTRODUCTION: TWO TIME CAPSULE TRIPS TO THE YEAR 2018

Following a ten-year battle to achieve uniform disposal of electronic waste in Europe, the framework of the WEEE Directive (Directive on Waste Electrical and Electronic Equipment) was adopted by the European Union in the spring of 2003. It should be noted that, although this framework is binding, it does not constitute national legislation in the individual member states of the European Union (E.U.). Following publication of the Directive in the *E.U. Gazette*, national states will have to pass appropriate legislation within 18 months.

It is already clear that, from 2005, it will be prohibited in Europe to simply dispose of electrical or electronic equipment together with domestic waste. All E.U. states will be required to have developed comprehensive product return systems by that date with manufacturers bearing the costs of recycling and disposal of non-recyclable residues.

The central aim of the Directive is to make manufacturers assume responsibility for their products. They will soon have to pay the costs for their treatment, recycling, and disposal. They must also ensure that a significant proportion of the products are re-used, or ensure that they are recycled according to best available practices. In turn the E.U. member states will promote the production of products which are easily dismantled and recycled (design for recycling). One should not forget, however, that to date (June 2003), the details have not been finalised by the European states.

At present, two fundamentally different interest groups are fighting for implementation. On group is calling for what we shall call a "monopolistic

Computers and the Environment: Understanding and Managing Their Impacts
Edited by Ruediger Kuehr and Eric Williams, pages 73–86.

consortium" in which, in principle, there would be pre-defined and pre-payable waste disposal costs for each piece of equipment produced in the future, with the level of total costs paid to be determined by the manufacturer's market share. The other group is calling for what we shall call a "market-driven system." Most European manufacturers have formed an (unusual) alliance based on common interests, and are demanding free competition between recycling companies; their aim is to promote new and innovative recycling methods. The proponents of the market-driven system believe that individual contracts between a manufacturer and the most appropriate recycling company will help keep costs at a minimum. They also believe that innovative products suitable for recycling (design for recycling) and new recycling technologies can significantly reduce costs and—in the final analysis—environmental damage.

At first glance, both models (the monopolistic consortium and the market-driven system) appear to be equally designed to meet the requirements of environmentally friendly disposal of electronic waste. Under closer observation, however, it becomes clear that there are dramatic differences between the two in terms of their effects on the environment, consumers, and industry. The authors invite you, the reader, to indulge in a little time travel. Come with us to the year 2018 to take a look at the possible consequences flowing from the two concepts and experience how the final choice of electronic waste disposal and recycling model will affect the environment and consumers over the next 15 years.

Almost all the new recycling technologies described during our two time-capsule trips are based on developments, research concepts, and tests that already exist today (in 2003). You'll find further information in the footnotes or in the reference section.

Follow us on our trip to the year 2018!

2. TIME CAPSULE 1: ECOLOGICAL PC RECYCLING (MARKET-DRIVEN SYSTEM)

EUROPE'S MOST MODERN RECYCLING CENTER OPENS

European Times, Hamburg, 14 February 2018: At a festive event yesterday attended by some of Europe's most prominent parliamentarians, Chancellor Angler declared Europe's most modern recycling center for used electronic equipment officially open. Ten million metric tons of discarded equipment—about two-thirds of Germany's entire output—will be processed every year in the new 400 million Euro Hamburg Recycling Center. Product recovery in the 800,000-square-meter facility is projected to be over 90 percent. The center was developed in close cooperation with most of the major electronic equipment manufacturers, who have signed long-term cooperative agreements to ensure its financial viability far into the future. Next to it stand two reprocessing facilities that will market the components recovered from the used equipment and distribute them through a newly established re-use distribution network.

2.1 A Tour through the New Recycling Center

The enormous floor space, with its complex of conveyor belt systems, looks more like a sterile production line than a scrap heap. (See Figure 1 for the flow of the overall recycling process.) At the end of the main conveyor, robots redirect the used equipment to a number of secondary conveyor belts that radiate out deeper into the building. Operators sit at control points and monitor the flow of used equipment. At many of the stations, workers remove components from the belt and direct them into nearby transport trucks. The steady flow of material is interrupted only by tunnel-like openings, where some of it is directed and automatically disassembled into pieces. After passing through most of the disassembly line, the only things left on the conveyors belts are blank circuit boards and mechanical parts.

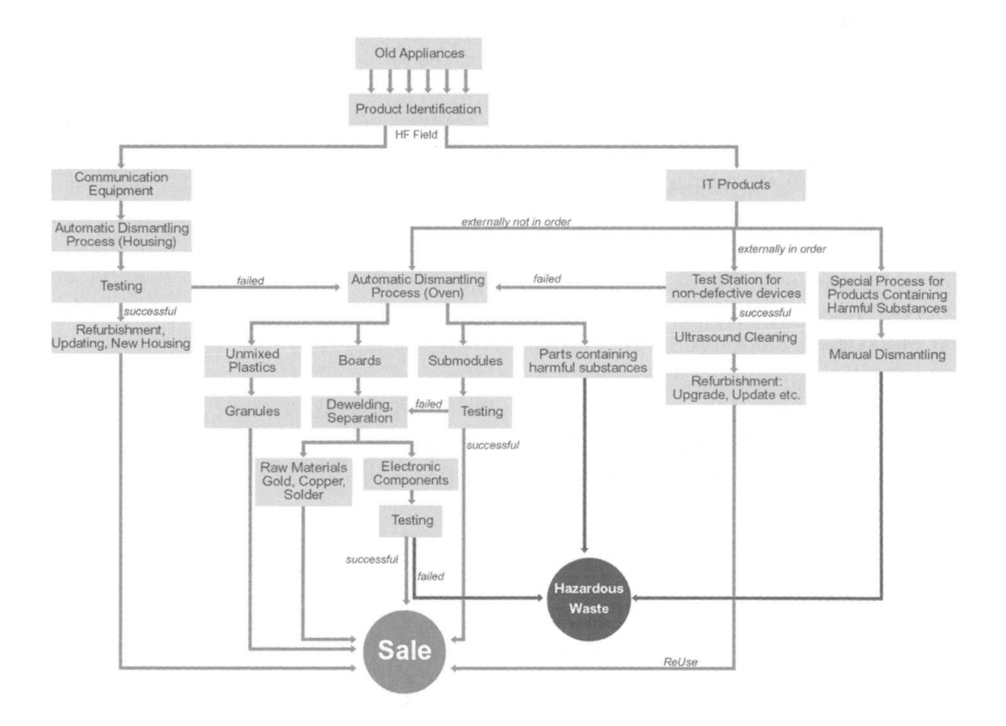

Figure 1: Recycling process in a modern recycling center in the year 2018.

Even the delivery of the used equipment to the plant is unusual. There are no bulk loads of discarded equipment dumped in piles in the yard. Instead, secured and bonded delivery trucks are directed to the Incoming Products bay, where workers carefully unload the old servers, personal computers (PCs), printers, and telecommunications equipment onto a wide conveyor belt. Spotting equipment that looks even slightly damaged is rare. "Strictly speaking, this is not scrap but valuable raw materials," said

Dr. Millar, director of the Hamburg Recycling Center. "After reprocessing in our plant, the units can very often be resold for several hundred Euro each."

"With new products and technologies constantly coming on stream—machines with more powerful processors and larger storage capacity, for instance—means that a lot of equipment is obsolete even as it reaches the marketplace. Most companies and consumers have gone through several computer generations in the last few years and written-off their old equipment long ago," according to Millar. He added, "Even simply for environmental reasons, it is a great plus that the manufacturers have decided to participate in this venture, even to the point of developing their products with recycling in mind right from the start" (Ruhr 2000).

Once the products are unloaded at the Incoming Products section of the facility, the equipment is moved onwards via the conveyor system to the actual recycling process area. The first station here—the most important element in the entire recycling process—is a one-meter-wide opening above the conveyor in which a strong, high-frequency radio electromagnetic (HF) field is generated. The HF field activates a tiny, integrated identity chip that is embedded in every item of equipment when manufactured. This ID chip contains data on product category, manufacturer, weight, and material composition. As well, it contains data gathered by special built-in sensors that monitored important parameters throughout the life of the equipment, such as the number of hours operated, temperature peaks, and voltage stresses (Figure 2).

The high frequency field generates a voltage on the chip to trigger the transfer of all of this information, which is then centrally collated and evaluated. The ID chip contains not only data referring to the equipment itself, but also data on sub-modules and components, known to be re-usable, that can be identified and indexed. The chip costs manufacturers only a few cents.

This technology—the activation of a microchip using a high frequency field without any physical contact—is a system that all manufacturers of electronic equipment agreed on using shortly after implementation of the original WEEE Directive in 2005.

Extremely old equipment, which may contain hazardous materials, is finally decommissioned at the end of its useful life and its components are sorted out. Hazardous materials are taken through a special disposal process—manual stripping down is often necessary. The costs for this additional and expensive stage for the removal of dangerous substances are separately invoiced to the original manufacturer. Automatic identification of the manufacturer facilitates precise allocation of higher return costs or recycling proceeds.

2.2 Visual Inspections of Equipment

Workers conducting visual inspections remove equipment that appears to be undamaged from the conveyor and direct it to an automatic testing station. As all units are equipped with a standardized test interface—test data are already known from the transponder system—a functional test is quickly and easily carried out. A positive result will direct the unit to an ultrasonic cleaning station and then onwards to a preparation stage for factory resale. Upgrade add-ons, firmware (software integrated in read-only memory [ROM]) updates, and new memory units transform a "scrap unit" into a ReUse product (Niesing 1999). Above all, in times of tighter economic times, these products are of interest for both household consumption and public services. For many rejected servers, PCs, and printers, this process secures a second or even a third life cycle (Federal Ministry for Education and Research 2003). This, however, represents only a small percentage of incoming used equipment, as advancements in the field of information technology (IT) are very rapid.

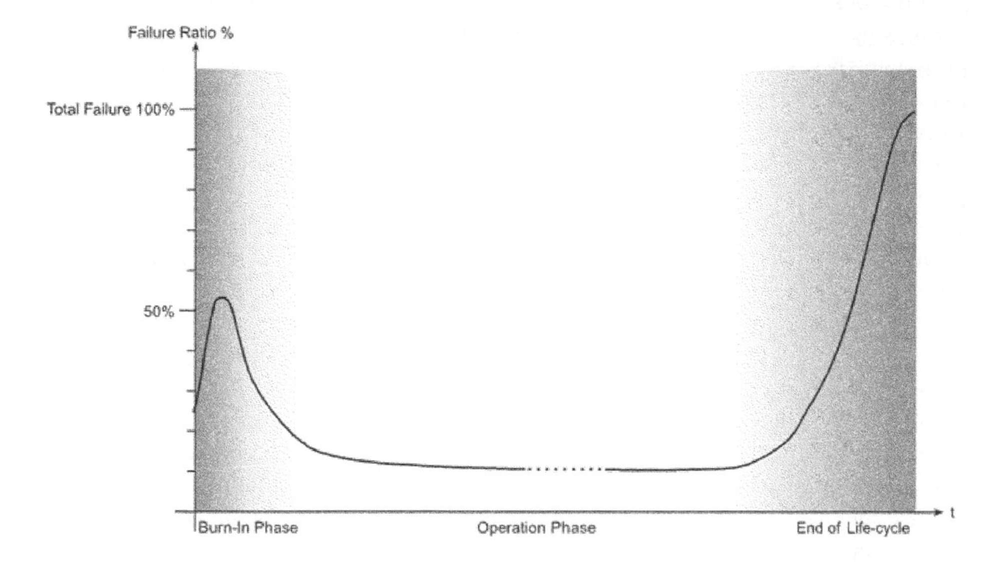

Figure 2: Failure ratio over lifetime of electronic components. Electronic components that have survived initial endurance tests and operating temperature peaks are put into operation, and are likely to operate smoothly for many years.

2.3 The Dismantling Process

All units that have not yet been sorted out travel along the conveyor system through an oven where the housings are heated up to a temperature of about 70 degrees Celsius. "Intelligent" metal clamps that hold the housing sections together open up at a pre-determined temperature and the housing quasi-opens up of its own accord (*Computer Partner* 2001).

The clamps are made of a metal with a "memory." This metal alloy may be shaped any way when cold, but when heated up it returns to its original form.

Several of these clamps react at different temperatures, thus making it possible to separate out the various components. Sub-modules, such as memory units, mechanical drives, and power supplies, travel on a conveyor to the next automatic testing station. Even with a faulty central unit, these modules are nearly always functional and re-usable. The stripped-down synthetic housings are treated with a water-soluble varnish, collected from the conveyor, and then sent on for further evaluation. At the next station, main boards and components are separated from each other in special "desoldering" ovens. An adhesive, with which the units are secured to the board, dissolves at a pre-determined temperature (Cornell University 2000).

Following the automatic unsoldering of processors, resistors, and capacitor chips, an automatic testing station checks that these components are still functional. These stations already know what components to expect (from the transponder system) and the appropriate test programs are started up.

Functional components and modules are sold through the sales office attached to the recycling center. "There is a ready market for used, functional, electronic components," says Dr. Millar. "Potential customers are, for example, manufacturers of gaming machines who prefer used parts. Older chip technologies are especially suitable for the manufacturer of white goods like washing machines or cookers. In addition to all of this, many of the recovered components end up in the retail trade as spare parts" (Breuning 2000). Memory chips are also used in the manufacture of toys, while fans and mechanical components are re-used for less demanding applications.

During the stripping-out process, critical components such as batteries are automatically removed from the housing. The lead-free solder is collected for further processing. Finally, with all the components removed, the printed circuit boards are all that remain, and although amounting to only 3 percent of a computer's total weight, they are present in all types of equipment in various forms (Gensch and Strubel 1999). The precious metals left on the boards, such as gold, silver, platinum, and palladium, are of particular interest because of their economic value. In a mechanical process, the boards are shredded and split up into isolated constituent groups in a number of different procedures. A hydro-metallurgical process isolates the metal components using different solvents and then they are recovered through extraction (Martin et al. 1997). This complex automatic process achieves recovery rates greater than 90 percent.

2.4 Recycling of Portable Communications Equipment

With the rapid rate of technological change, communications equipment such as mobile phones, personal digital assistants (PDAs), and other portable communications equipment become obsolete after two to three years. In general, however, the equipment is still operational, even after the model has been superceded by a new one. After a few years, very often these products are regarded as out-of-date from an aesthetic point of view, or else their internal power sources may have reached the end of their life. Rarely does one find a unit with mechanical damage on the special disassembly conveyor.

Here, as well, the system recognizes all the basic parameters from the integrated transponder chip. The housing covers open up automatically, but in this case the boards are not split up to the component level but are fed onwards to a testing station. This controls the next processing stages. Faulty boards are fed to one of the desoldering stations described above, but further processing is carried out for still-functional components. For circuit boards classified as being still functional, the energy cells are replaced, software is updated, the boards are provided with a fashionable new casing, and the unit is returned to the market as a reprocessed product. In this way, the communications products, previously classified as obsolete, are given a new lease on life with technical innovations integrated via software, while the cases are changed to satisfy market trends. This type of reprocessing, which could be considered more as a product update, is carried out in close cooperation with manufacturers, who supply the new cases and software updates. For these "new/old" communications products, a second market has evolved, which, in spite of the enormous flow of new equipment, uses these products a second or third time.

2.5 Summing Up—Recycling Under the Market-Driven Approach

Profits arising from resale are transferred back to the manufacturer and reduce the general recovery cost. Analysis of reprocessing processes is fed back directly as part of the design considerations for new equipment (Hesselbach et al. 2000).

The modern management team in charge of the Hamburg Recycling Center is concerned with constant improvements in efficiency and cost reductions—all based on high quality standards. The team recognizes that it faces tough competition from other recycling facilities, as manufacturers are entitled at any time to sign up with the product recovery partner of their choice.

This recycling center of the year 2018 is not a scrap recovery company but rather a sophisticated fabrication and marketing operation. It not only processes electronic scrap for resale with its own production line, but it also defines the costs of recycling in very close cooperation with the electronic equipment manufacturers. By sharing in the proceeds of recovery, the operation is interested in achieving the lowest level of complexity for disassembly and the highest resale value for the basic materials, components, and reprocessed units (Behrendt et al. 1999).

3. TIME CAPSULE 2: PCS BECOME ELECTRONIC JUNK (MONOPOLISTIC CONSORTIUM SYSTEM)

ELECTRONIC WASTE DISPOSAL COSTS RISING—WEEE Consortium does not rule out price increase for disposing of old equipment.

European Times Hamburg, February 2018: The consortium anticipates that the year 2018 will see a volume of over 15 million metric tonnes of retired electrical equipment. "Despite state-of-the-art dismantling technologies, manual dismantling costs have continued to rise. We find ourselves forced to pass these costs on to the manufacturers," according to Edward Dustin, executive director of the WEEE

Consortium, speaking at a press conference following the AGM which brought together the Consortium's members. "Although it is true that the recycling ratio is almost 20 percent, the proportion which cannot be recycled has to be stored in specially designed depots—a costly procedure which has to be carried out in as environmentally friendly a manner as possible. This, too, entails higher costs for us." Initial estimates predict that IT equipment manufacturers will see their costs increase by 3 to 4 percent. If the manufacturers' reaction is anything to go by, this increase will be reflected in the equipment's retail price.

3.1 A Tour through the System

Electronic scrap is roughly trundled from trucks into a deep, scrap-filled bunker—PCs, video display units (VDUs), servers—all completely unsorted. Products with no visually discernible defects or damage are crushed into junk by the weight of the units above. A conveyor belt transports this desolate jumble of indefinable metal, components of synthetic materials, plastics, printed circuit boards, and electronic components to the hazardous materials disposal bays. Earlier, at the local authority collection points, the scrap was roughly pre-sorted into equipment groups in containers—large and small white goods, brown goods, entertainment units, IT equipment, and office and communications equipment—with no concern for reclamation or reprocessing (Hauser and Roettchen 1995).

Regardless of whether the unit was the product of an optimum "design for recycling" production process or the manufacturer had avoided the use of hazardous substances in its production, all equipment ends up in the same waste disposal process. The chaotic mixture of screws and clamp fixings, along with insoluble adhesives, inhibits any meaningful recovery process.

Large magnets separate the metal housings with electronic components inside from plastic cases and place them onto a conveyor belt. Here, workers armed with electric shears remove all non-metallic parts from the casings. The metal sheeting is collected and the printed circuit boards—complete with electronic components, motors, and power supplies—are placed onto a conveyor for shredding.

The electronic components inside the cases made of synthetic material likewise end up on the conveyor for shredding, neither stripped down nor further processed. The giant shredder chops up this mix of plastics, printed circuit boards, and component scraps into tiny pieces; magnets separate out the metals from the lighter, shredded materials, and a float-or-sink process stage sorts the synthetic substances according to specific weight. The majority of the electronic scrap, however, cannot be cleanly separated out, and must be burned in the large waste incineration plant attached to the recycling center.

Even for the synthetic materials, which have been sorted by type from amongst the electronic scrap, remarketing has become ever more difficult, despite the incentive of increasing oil prices. Over time, the larger plastics manufacturers have moved completely out of the recycling field, because of a lack of raw recyclate material available on the market. They are hardly likely to become interested in new products.

The consequences for electronic scrap recycling are quite clear: nobody feels in any way responsible for the clean dismantling of plastic cases or for their sorting, handling, and resale.

Electronic scrap is likewise turned into fodder for the shredder. With the wide variety of materials, which includes additives such as metal oxides, cadmium, and chlorinated anti-flammable coatings, the shredded output presents problems that makes these materials almost impossible to recycle. All attempts to separate and sort the material are expensive—costs which are not recoverable on the used-product market. Even for what the manufacturers declare to be "unmixed plastics" there is no incentive to separate them out, as expensive sorting methods such as those using infrared or X-ray fluorescence are out of the question. The various other licensed waste disposal options—dumping in a landfill, incineration, or using as aggregate by the construction industry—inhibit any positive initiatives and pose serious threats to the surrounding communities, polluting surface and ground water over the long-term and causing global environmental problems (Faller et al. 2001). Incinerating the untreated electronic scrap produces and disperses hazardous substances into the environment—but this practice has continued more or less unaltered for years.

4. TWO SCENARIOS—WHY SO DIFFERENT?

In the second scenario described above, the WEEE Directive is implemented by a monolithic consortium that sets the general waste disposal prices for each product within a product category. The waste disposal costs for each individual item of equipment that ends up in the "destruction" process are not calculated; the costs are allocated according to a manufacturer's market share of all equipment of that product category. This means that in the future, disposal costs must be paid in advance—regardless of whether, how, or when a product will be discarded. This approach represents a type of contract between generations—with an uncertain outcome. On the surface, this approach does not appear to be economically viable on its own. In this scenario the manufacturer has no motivation to optimize his equipment or product design for recycling or to invest in recycling technologies.

As the costs for disposal in the case of the "throw-away" product philosophy are divided amongst all manufacturers, individual manufacturers receive no feedback when savings are made through their new recycling methods. Higher investments in designing products for recycling (such as using thermal clamps on casings for easy opening) produce no financial return. It is impossible to switch to another recycling system; the consortium's monopoly position prevents this. Therefore a manufacturer has no interest in investing in the development of products that take design for recycling as an important feature. What motivation, for example, would a manufacturer have to avoid cheap materials when producing a product?

Under the second scenario, innovative manufacturers could even be punished twice, because their higher production costs are not matched with cost reductions for disposing of their recycling-friendly products. A competitor, however, who has not invested in design for recycling, profits from the low average price achieved for all corresponding products on the market, although he might be the only one who has not invested in recycling-friendly design. Actually, the recycling costs of his own products should be higher than those of his competitors. Thus, no motivation exists for a manufacturer to be involved in innovative technologies, long product life, recoverability, or pioneering ideas in recycling methods.

There is another negative consequence of this system: manufacturers have no interest in becoming involved in any recovery consortia, so other interest groups have stepped into the vacuum. The result: material recovery costs are not transparent, and a bureaucracy of administrators, managers, and consortium members must be paid, adding to the cost of the overall system. A horror scene! Negative examples of practices such as these can already be observed in some European countries in 2003.

In our first scenario, the national legislators put the power of the free market to work for implementing the WEEE Directive. Different take-back systems compete with each other. Waste recovery enterprises such as the one described in our example offer their services, as do other concerns that return used equipment to the market by completely different processes. The enterprises determine their individual recycling costs of each disposed product per manufacturer on their own. For example, recycling company "A" disposes the PC of manufacturer "X" at a price of €3, whereas recycling company "B" offers the disposal of the same PC at only €2. The manufacturer, in turn, concludes medium- or long-term contracts with the most acceptable recycling processor. Costs arise only when a product is fed into the recycling process. Competition between the various reprocessing systems demands the same quality standards and efficiency improvements from the participants in the recycling business, just as in other production undertakings. Because the exact waste recovery costs are allocated to the specific product of each particular manufacturer, motivation to design for recycling is assured. If a manufacturer invests in a development that facilitates quick and easy dismantling of a product (such as thermal clamps), the resulting cost savings are credited directly back to the manufacturer (Niewoehner and Renz 2000). Also, because profits arising from resale are transferred back to the specific manufacturer, some of them rethink their business models and manufacture long-life products where a market for such products exists.

The future outcome of whether or not products will be manufactured for easy recycling—thereby reducing costs for consumers—depends not only on environmental legislation but also on the organizational formation of the recycling process.

5. OPTIONS IN 2003

The WEEE Directive came into effect in February 2003, but what does this mean for European governments? If not already the case, it means that every country of the European Union must pass certain national legislation by 1 August 2004. The directive is, however, mostly just a description of a conceptual framework that defines the room in which industry has to maneuver. It remains open whether implementation of the legislation will be via a consortium or via independent service providers. The parliamentary process to introduce such legislation generally takes about a year—hearings and one to three readings in parliament take that much time.

5.1 National Legislation Process—An Example

Germany is one of the countries that is just starting the process of passing legislation to comply with the WEEE Directive. Using it as an example, the authors briefly sketch the process.

Phase 1 (Unofficial Process)

1. Discussions are launched between the relevant ministries and associations. Germany was in this phase in April 2003. The federal Ministry for Environment, Nature Protection and Reactor Safety and the Ministry for Economics have initiated discussions on WEEE implementation with associations like the ZVEI[1] and BITCOM.[2]
2. These discussions lead to a "green paper," the basis for internal discussions among ministries, which details the various feedback and ideas of the civil servants within each ministry.
3. After internal exchange of opinion among the various ministerial working groups, selected external interest groups are invited to discuss the proposals (still unofficially).
4. These discussions lead to the publication of unofficial draft legislation, which is presented to everyone involved thus far.

Phase 2 (Official Process)

1. Compiling the results of official discussions and modifications, staff convert the unofficial draft legislation into the official draft legislation, which is then presented to cabinet.
2. With the government's approval, the draft is laid before parliament.
3. This parliamentary process eventually leads to an official bulletin announcing the new legislation.

---

[1] Central Association for the Electrical and Electronic Industries, reg. ass., Germany.

[2] Central Association for the Information Industry, Telecommunications and New Media, reg. ass., Germany.

5.2 Different Interest Groups—Who Supports Which Solution?

5.2.1 The Monopolistic Approach

In many European countries, association representatives clearly support the idea of a consortium, with aspirations to set up a monopoly which they can control, and whose services they can offer their members. The self-interests of those involved must be kept in mind, however, because monopolistic organizations offer highly paid directorships and positions along with executive powers. Examples of this situation were considered in Belgium and Austria (with decisions expected).

More than a few companies support the monopoly approach. There are a number of reasons for this. Often in these companies, personnel who are in charge of environmental issues are brought into the decision-making process—but often they are not business-oriented and overlook the long-term financial consequences of decisions for the company. At the same time, some manufacturers control one association or another and thereby support the monopolistic viewpoint.

The varying internal positions in different companies are surprising. A local manager of an organization operating on a Europe-wide basis, who is responsible for the company in his own country, frequently supports the monopolistic approach, as this for him is the simplest approach—no problems with associations or officials, no individual positions competing with others, no individual creativity—merely the takeover of existing structures. On the other hand, a responsible, European-oriented directorate in the same organization is more likely to recognize the long-term consequences and, above all, the overall costs of a monopolistic approach.

Companies whose products are manufactured in a less environmentally-friendly manner are naturally more interested in a monopolistic system, because the costs of waste disposal and recycling are uniformly distributed among all manufacturers, irrespective of whether or not their products have been produced with environmentally-friendly designs.

5.2.2 The Market-Driven Approach

In general, it can be concluded that forward-looking organizations view environmentally-sound product development as a fundamental element of their future product portfolios. They want to establish a system in which they themselves have control over waste disposal. As of April 2003, the companies in Germany that had indicated support for a market-driven solution included almost the entire German IT industry, about one-third of the consumer electronics industry, and almost all the worldwide, market-oriented players.

In Germany, the competitive system appears likely win, although by the summer of 2003 it still remained open as to who would control it.

6. CONCLUSIONS FROM TWO SCENARIOS FOR 2018

For a fair, sustainable, and environmentally-sound conversion and implementation of the WEEE Directive, the costs of recycling electronic equipment must be directly allocated according to the "polluter pays" principle. Those manufacturers who are environmentally aware and conscious of their responsibilities should be rewarded, and in the final analysis, will produce goods more economically. In this way consumer pressure would force the less progressive manufacturers to change.

This reality is only possible, however, in a system where the product-return and recycling costs are not simply allocated to a particular product or product group, but rather, imposed on individual manufacturers based on the actual costs arising from the recycling process of their product. A recycling system based on a monopoly does not allow such allocation; it inhibits cost-efficient recycling that could arise through competition between product-return systems, and also reduces the incentive to develop innovative recycling technologies.

European society should see the WEEE Directive not only as a challenge but also recognize it as a great opportunity for themselves and the environment.

REFERENCES

Behrendt, Siegfried, Ralf Pfitzner and Rolf Kreibich. 1999. *Competitive advantages through ecological service provisions: Conversion in company practice.* Berlin: Springer Verlag.

Breuning, Ralf. 2000. On high value levels. On the basis of decision criteria, possibilities and limitations can be shown in the renewed use of electrical/electronic units. *Muellmagazin* (Waste Magazine) 13(4): 23–29.

Computer Partner. 2001. Intelligent metals replace plastic fasteners. *Computer Partner*, 1 October. <http://www.golem.de/0110/16106.html> (18 March 2003).

Cornell University. 2000. Adhesive breakdown to recycle old computers. Cornell University news release, 22 August 2000. <http://www.newswise.com/articles/2000/8/ATERP.CNS.html> (18 March 2003).

Faller, Hubert, Gerhard Ott and Ulrich Wurster. 2001. Hazardous materials pollution in electrical / electronic recycling—A work protection problem? (Exposure to hazardous substances in the recycling of old electrical/electronic devices). Gefahrstoffe - Reinhaltung der Luft (Hazardous Materials—Maintenance of clean air [Air Quality Control]) 61(5): 175–179.

Federal Ministry for Education and Research. 2003. Verbundprojekt ReUse (Cooperative ReUse Project). Federal Ministry for Education and Research, Tech. University Berlin, Kubus (Cooperation and Advice Center for Environmental Issues). <http://www.tu-berlin.de/zek/kubus/Reuse/reuse.html> (18 March 2003).

Gensch, Carl-Otto and Volker Strubel. 1999. Insights with prospects. Manufacturers develop environmentally friendly technologies for electronic products. *Muellmagazin* (Waste Magazine) 12(4): 55–60.

Hauser, Henrik and Peter Roettchen. 1995. Electrical and electronic equipment. *Waste disposal logistics III: Circulation economics*: 139–160. Berlin, Bielefeld, Muenchen: E. Schmidt.

Hesselbach, J., C. Herrmann and H.-Y. Kang. 2000: Recycling oriented product development—Evaluation and solutions science. In *VDI reports 1570, Recycling orientated development of technical products 2000: Management of complex target conflicts*, 159–174. Duesseldorf, Berlin: VDI-Verlag.

Martin, Rainer, Klaus Niemann and J. O. Funk. 1997. Exploitation of electrical and electronic equipment. *Kunststoffe* (Synthetic Materials) 87(12): 1744–1746.

Niesing, Birgit. 1999. Products live longer. Environment Technology. Fraunhofer Magazin (Fraunhofer Magazine) (2): 40–42.

Niewoehner, J. and R. Renz. 2000. Visual registration for the quantification of dismantling times of old electrical/electronic equipment. In VDI Reports 1570: Recyclingorientierte Entwicklung technischer Produkte 2000: Management komplexer Zielkonflikte (Recycling orientated developments of technical products 2000: Management of complex target conflicts), 51–65. Duesseldorf, Berlin: VDI-Verlag

Ruhr, Paul-Albert. 2000. Kostenueberwaelzung im Rahmen der Produktverantwortung aus der Sicht der Industrie (Cost transfers within the framework of product liability from the viewpoint of industry). Kosten in der Abfallwirtschaft: Druckschrift zu den 8. Koelner Abfalltagen (Economics of waste. Proceedings of the 8th Conference on Waste) 3–4 November 1999, at Cologne, 397–406.

Chapter 5

IBM'S ENVIRONMENTAL MANAGEMENT OF PRODUCT ASPECTS

Reinhard Höhn[a] and Anne Brinkley[b]

[a]*IBM, Germany*
[b]*IBM, USA*

1. INTRODUCTION

The International Business Machines Corporation (IBM) strives to lead in the creation, development, and manufacture of the industry's most advanced information technologies, including computer systems, software, networking systems, storage devices, and microelectronics. The company's worldwide network of IBM solutions and service professionals translates these advanced technologies into business value for its customers.

IBM is the world's largest information technology company, employing approximately 320,000 employees worldwide, with operations in over 160 countries. Revenue for 2002 was U.S.$81.2 billion. The company's major business segments include information technology hardware, software, services, and global financing.

IBM is committed to environmental leadership in all of its business activities, from its operations to the design of its products and use of its technology. The company first formalized this commitment in its Corporate Policy on Environmental Protection in 1971. IBM's global environmental management system supports that policy and, through this system, the company implements programs to achieve its objectives. Because of its global environmental management system, IBM earned a single global registration to the ISO 14001 environmental management standard in 1997. This registration covers the company's manufacturing, hardware development, and research operations worldwide.

The following provides insight into IBM's approach to manage environmental aspects of its products in a fast-changing business environment and to reduce the environmental footprint of the businesses conducted by IBM worldwide.

Computers and the Environment: Understanding and Managing Their Impacts
Edited by Ruediger Kuehr and Eric Williams, pages 87–98.

The potential environmental impacts of electronics products are diverse, and the collective assessment of them (by producers and consumers of these products) is a continuous exercise. Just as these impacts of products are influenced by the dynamics of design, economics, new technologies, and waste management alternatives, the effective management of the environmental impacts of products must also be a dynamic process that responds to, as well as affects, changes in environmental aspects of production operations. IBM successfully manages these dynamics through the integration of product design processes within its corporate environmental management system.

In 1990, objectives with a product focus were added to IBM's corporate policy on environmental affairs, because the ongoing internal as well as external discussion expanded its focus on process-related issues to include product-related issues. Customers, especially from Europe, made special requests for environmentally conscious products. These new objectives directed the company to take the following initiatives:

- Develop, manufacture, and market products that are safe for their intended use, efficient in their use of energy, protective of the environment, and that can be re-used, recycled, or disposed of safely.
- Utilize IBM products, services, and expertise around the world to assist in the development of solutions to environmental problems.

Integration of these policy objectives throughout the corporation was accomplished using the core elements of IBM's environmental management system; that is, program requirements were established for product environmental objectives through three key management tools of the system: (1) corporate instructions, (2) definition of significant aspects and corporate metrics, and (3) standardized design practices (described below).

2. CORPORATE INSTRUCTION

IBM formalized a worldwide program for the development of environmentally conscious products through the preparation of a corporate instruction. Corporate instructions in IBM are *global* directives used to support and implement the company's policies. Compliance is required by all affected business units. The *Environmentally Conscious Products* corporate instruction defined specific objectives for the design of all hardware products. These objectives addressed the necessity of design considerations for the following:

- Product re-use and recyclability at the end of the product's life
- Upgradeability to extend the product's life
- Use of recycled materials where economically and technically justifiable
- Improvement in energy efficiency
- Safe disposal at the end of the product's life
- Selection of environmentally preferred materials and finishes to minimize resource use and environmental impacts

Further, it established an organizational infrastructure accountable for implementation of product environmental objectives in every operating division that had product design responsibilities. An executive strategy owner for each hardware division was identified to execute implementation of product design requirements and report the division's strategies and annual performance against the specific design goals identified by the division for its product sets.

3. SIGNIFICANT ASPECTS AND CORPORATE METRICS

The Corporate Instruction on Environmentally Conscious Products explained how divisions would manage program objectives for environmental design. What they manage—the definition and prioritisation of significant product environmental aspects—was fundamentally defined at a corporate level in the environmental management system for reference by IBM's hardware divisions. A team of internal experts identified a list of product attributes and environmental aspects that apply to a majority of the products developed and marketed by IBM. The determination of significant environmental aspects is based on these experts' consideration of the environmental impact of a particular aspect, legal requirements, IBM environmental requirements, and customer views. IBM's corporate environmental management system lists the following significant aspects for products:

- Product energy consumption (direct energy consumption, heating and cooling requirements)
- Product size and weight
- Product chemical emissions
- Potentially hazardous parts and assemblies (e.g., batteries)
- Product protective packaging (for shipping)
- Product supplies and/or consumables (e.g., toners, cartridges, paper, storage media)
- Re-use and/or recyclability (design for disassembly, product return, re-use, upgradeability, and/or modularity)

These categories of aspects have been consistent throughout annual reviews of the environmental management system. New aspects may be proposed by divisions, by corporate staff, or by members of IBM's worldwide environmental community for consideration. In addition, corporate environmental staff members establish targets or metrics for corporate-wide performance to bring emphasis to product aspects of prime importance. Division strategies for product management must align with these corporate metrics and provide the resources to implement them. For example, to support the corporate metrics on use of recycled plastics, the hardware product divisions have to submit a plan for their estimated annual use of recycled content plastic versus the total amount of plastic that will be used annually for all plastic parts production for the division's product lines.

Corporate metrics for product aspects have been part of IBM's environmental management system since 1996. Metrics requirements are established by corporate staff with input from the operating divisions to reflect the dynamics of technologies, customer requirements, and corporate leadership initiatives. The metrics are reviewed at least annually with a committee of IBM's board of directors. Since the inception of the metrics, targets for the IBM Corporation have reflected the corporation's commitment to voluntary programs like ENERGY STAR®[1] development and increased use of recycled plastics in product applications, the assessment of products for implementation of comprehensive design-for-environment practices, and reduction in landfill use for product end-of-life management.

IBM's personal computer products illustrate an evolution of designs that have led to changes in the environmental aspects of the product lines and in specific corporate metrics. For example, in the early 1990s IBM implemented a successful closed-loop plastics recycling operation for keyboard manufacturing operations in Scotland to demonstrate the technical and economic viability of recycled plastic for PC applications. The technical success of the program gave feasibility to a corporate metric for conservation of materials resources through recycled content plastics use. The closed loop program was discontinued, however, when the resin system, poly vinyl chloride (PVC), was replaced due to customer preferences, and alternative resins were not cost-competitive in a recycled form. Nevertheless, IBM continued to focus on recycled plastics usage for PCs, since high volumes of limited plastics types were required for PC production for both consumer and commercial product lines. A corporate metric to increase use of recycled plastics began with modest targets of 2 percent growth per year as IBM pursued commercial sources of cost-competitive, recycled content plastics from the plastics industry to address this product aspect corporate-wide. Through development partnerships with plastics compounders, IBM succeeded in 2000 in implementing a 100-percent recycled content polycarbonate alloy for all the major plastic parts of the IBM IntelliStation E Pro system unit, a high-end desktop PC. This innovation ultimately led to the widespread use of recycled content plastics in most IBM product lines, including all desktop PCs, surpassing corporate targets that had reached 10 percent of all plastics used annually by IBM in worldwide production. But IBM's PC business was changing course. The advent of IBM's e-business strategies turned the company's attention to business customers and IBM withdrew from the consumer market for PCs. The products' design criteria changed and introduced different environmental impacts.

Today's IBM IntelliStation E Pro employs a metal housing, which commercial customers preferred for durability and cost issues; the design of plastic housing needs an additional metal liner for electro-magnetic shielding, which can be avoided with pure metal housings. The transition of other IBM desktop brands like NetVista to more metal construction reduced environmental impacts related to plastics life cycles and brought about greater focus on the environmental aspects of decorative painting of

[1] ENERGY STAR® is a U.S. registered trademark.

metal chassis. Corporate metrics were subsequently established for increased use of powder coatings that don't use solvent in order to eliminate volatile organic air emissions and solid waste from overspray during fabrication of metal housings.

4. STANDARDIZED DESIGN PRACTICES

The third element of the management system for products at IBM is the Corporate Standard for Environmentally Conscious Design. The standard is owned and maintained by the IBM corporate program manager for Design for Environment and is released through a process controlled by the corporate standards community within IBM. Contents of the standard address legal and IBM internal requirements for hardware products, ranging from banned and preferred materials to energy consumption, design practices for disassembly and recyclability, upgradeability, labelling of plastics and parts that require special handling at end-of-life, battery management, and product take-back, among other requirements. The revision process for the corporate standard includes review by division level and worldwide environmental staff. Country-specific requirements, especially regulations, can be included in the corporate standard as a reference document known as a National Bulletin. Compliance to the corporate standard is mandatory and is certified through product development processes that must be completed by each division prior to the general announcement of new products. The Corporate Standard for Environmentally Conscious Design ensures that IBM's environmental objectives for product design are met uniformly throughout the company even in the absence of comparable environmental regulations for products in some markets.

As IBM does not manufacture all products or parts and components by itself, design requirements have a direct impact on the supply chain. Without good cooperation with suppliers, the overall goals for product environmental attributes cannot be achieved. IBM's design requirements are communicated to suppliers through a corporate engineering specification which concisely provides essential requirements without extraneous references to internal procedures. The engineering specification applies to all materials, parts, and products procured for IBM applications.

5. EXTERNAL INITIATIVES

IBM recognizes that internal activities governing the management of product environmental aspects must be part of a bigger solution for global environmental management. To that end, IBM products, services, and expertise have driven external initiatives to establish and promote international programs that advance effective and environmentally sound practices for the design, marketing, and end-of-life management of information technology products. Examples include the internationalization of the U.S. ENERGY STAR® program, the development of standardized, voluntary environmental declarations for product attributes, international design standards for environmental considerations for electronic products, and extensive product recycling services, as outlined below.

5.1 The International ENERGY STAR® program

As a charter member of the ENERGY STAR® program initiated by the U.S. Environmental Protection Agency[2] in the early 1990s (see Chapter 8 for introduction to ENERGY STAR), IBM expertise supported the development of appropriate performance targets and technologies for energy-efficient office equipment, such as computers and printers, to reduce the environmental impacts associated with electricity consumption during their use phase. IBM's uninterrupted participation in the program and consistent adoption of its program criteria for corporate metrics in the area of product energy efficiency attest to the value that IBM has placed on this voluntary program. Having become a de facto requirement for sales of office equipment to the U.S. government, ENERGY STAR® recognition among global PC manufacturers was high. As similar programs emerged from national energy agencies in Europe, IBM promoted internationalisation of the ENERGY STAR® program in both Europe and Japan. Members of IBM's worldwide environmental staff worked to promote and establish these international ENERGY STAR® programs, which have, in turn, influenced energy efficiency criteria in many of the eco-labelling programs for PCs in these regions.

5.2 Environmental Standards for Electronic Products: ECMA-341

The perception of PC design and manufacturing as an "in-house" enterprise is incorrect. Many integral parts of a PC system are purchased as commodities from a global supply chain. A computer monitor may come from one contract manufacturer, the keyboard from another, and the system unit possibly from another. While environmental design requirements may be well recognized and managed for in-house operations of a company, there is a lack of harmonization of environmental specifications from different manufacturers to guide their common suppliers. IBM has sought to provide standardization of a set of basic and sound environmental design practices for the industry as a whole through the development of an international standard through ECMA. ECMA International, the former European Computer Manufacturing Association, is based in Geneva, Switzerland and serves as a standards organization for information and communications systems. In 2001 and 2002, IBM led a technical working group, TC 38, within ECMA to draft a set of requirements for electronic product attributes that demonstrate environmentally conscious design.

[2] ENERGY STAR® program by the U.S. Environmental Protection Agency <http://www.energystar.gov> (10 April 2003).

ECMA-341, "Environmental Design Considerations for Electronic Products,"[3] was approved and released in January 2003. In addition to mandatory criteria, the standard includes numerous design recommendations and offers a model design checklist to assist companies in developing their own management tools for integration of the mandatory and suggested environmental aspects into product design. Design requirement areas covered by this standard include the following:

- Material efficiency
- Energy efficiency
- Consumables and batteries
- Chemical emissions
- Extension of product lifetime
- End of life
- Substances and preparations needing special attention

Standardization of environmental design requirements provides a focal point for consensus between multiple manufacturers and their suppliers to address common goals and jointly influence the magnitude of specific environmental impacts. Adoption of standards criteria can reduce business costs, because common practices and expectations can be readily addressed by suppliers. Finally, standardization of design criteria helps to define environmental aspects and their assessment in a uniform way. This ultimately benefits both manufacturers and consumers in that meaningful, ethical, and comparative communications about product environmental aspects can be promoted in the marketplace to support informed decision-making by consumers.

6. SUPPLY CHAIN MANAGEMENT

Another example of IBM's desire to support international initiatives and reduce the inefficiencies of sole national approaches to supply chain management is the idea to harmonize formats and lists of reportable substances in the information and communication technology (ICT) industry. A reportable substance is a chemical that is regulated by legislation or associated with issues of sustainability. This latter category includes substances that have environmental, health, and safety interest (e.g., polyvinyl chloride) or economic interest for recyclers (e.g., silver, copper, gold, etc.) The monitoring of such substances by the supply chain varies significantly with unique market influences in different geographies. Recently, the European Information, Communications and Consumer Electronics Industry Technology Association (EICTA)[4] initiated an effort to maintain a list of reportable substances for common reference by manufacturers in their supply chain management to overcome the burden on suppliers of answering various questionnaires and specifications used by different

---

[3] "Environmental Design Considerations for Electronic Products," ECMA Standard ECMA-341, Geneva, January 2003, <http://www.ecma-international.org/publications/standards/ecma-341.htm> (10 April 2003).

[4] <http://www.EICTA.org>

manufacturers. Similar activities had also begun in industry groups in Japan and the United States. IBM immediately saw the need to combine these efforts and hosted initial meetings of all three groups to achieve an international approach. In 2003, EICTA, the U.S. Electronic Industry Alliance (EIA), and the Japanese Green Procurement Survey Standardization Initiative (JGPSSI), with the support of IBM, are working to develop a uniform template for substance and material declaration using common criteria for their identification and reporting. The aim is to standardize data gathering throughout the global supply chain by providing a common format for structured materials data exchange. In addition, the process of standardization will establish a set of rules that govern the selection of substances necessary for disclosure, as well as a consensus procedure for modifying the substance lists.

A global standard for materials declarations will facilitate supply chain management for product environmental improvements by reducing the cost of data collection as well as the reporting burden on suppliers. A common format is also helpful for the final customer, who might request substance reports from several manufacturers and makes it easy to integrate information from different sources.

7. DECLARATION OF ENVIRONMENTAL PRODUCT ATTRIBUTES: ECMA TR/70

The voluntary declaration to customers of product environmental attributes of information technology products was also an effort undertaken by IBM through ECMA's technical committee from 1995 on.

As customers heeded the recommendations of various programs designed to identify environmentally preferable products, such as national environmental labelling schemes, again standardization issues detracted from manufacturers' participation. Eco-labels (e.g., German Blue Angel, the Nordic Swan, or the Japanese Eco Mark) are only awarded if all requirements of a specific scheme are fulfilled, but cannot differentiate further. Also, these programs applied to only a small range of IT products (e.g., PCs, notebooks, or printers up to a certain printing speed).

To foster a common reporting format and give customers the opportunity to review a full range of environmental aspects related to products from each manufacturer from whom they could buy, IBM and other manufacturers in ECMA developed a common declaration scheme for product related environmental aspects. A technical report, TR/70,[5] available free of charge from the ECMA Web site, lists the aspects to be declared, describes and references the standards which have to be followed if product aspects need measurement, and suggests a format which follows the international standard EN 45014 for declarations of conformity. Examples for selected IBM

[5] "Product-related environmental attributes," ECMA Technical Report TR/70, second edition, Geneva June 1999.

products can be downloaded from IBM environmental Web sites (e.g., <http://www-5.ibm.com/de/umwelt/>).

8. PRODUCT TAKE-BACK

The potential environmental impacts of electronic product disposal are among the most prominent issues in the information technology sector and the impetus for product recycling legislation in many of the globe's major markets. These impacts are influenced by factors beyond a sometimes singular focus on product composition. Product volumes, regional recycling infrastructures, reverse logistics strategies, and the extent of customer response to product return offerings contribute to the complex challenges of product end-of-life management.

IBM has been proactive in addressing each of these contributors to potential environmental impacts of product disposal. Apart from the company's design focus to minimize the use of materials which require special handling at end-of-life, IBM has invested in a number of other business operations that demonstrate leadership in reducing or containing product waste at end-of-life. These operations include leasing and remanufacturing programs, demanufacturing operations, and used-parts recovery (useful for manufacturing new products and harvesting potential spare parts), product take-back services, and a global array of product end-of-life management (PELM) centers which process tens of thousands of metric tons of retired equipment annually.

All these aspects of product end-of-life management begin with the return of products to one of over seventy PELM centers in IBM's network of facilities around the world. The returns can be classified as products used internally at IBM, end-of-lease returns, returns owned by business customers, and returns from private or small customers.

All these returns come in different condition, resulting in different operations and return rates, and needing different solutions.

IBM's capabilities in product recoveries were founded in the company's early business model for equipment leasing. Leasing provided incentives for recovery and refurbishing or upgrading of equipment which would retain its value for extended revenues. Also, in-house parts recovery from returned machines provided inventory for warranty and field repairs. Leasing programs typically deal with later model equipment which has a high probability of retaining value for remanufacturing upon return to IBM. Refurbishing operations belong to IBM's Global Financing division, which remarkets the used equipment. Equipment that is returned from lease programs arrives first at IBM refurbishing centers, which then determine the product's capacity for re-use. If the product is determined to be re-usable, it will be cleaned, repaired, and upgraded as needed. After thorough testing, the product is resold by IBM Global Financing directly or through broker partners.

Equipment that is not salvageable for machine resale is directed to demanufacturing centers for parts recovery and materials recycling. This equipment originates from

IBM internal assets (used by the company) and from customer-owned equipment returned through formal and informal product take-back services available to IBM customers.

The internal IBM returns are governed by company procedures requiring that every internally-used IT product has to be shipped to a PELM location for disposition. Delivery to other internal IBM functions who might have a need for that specific product is enabled through a Web-based offering tool. Internal returns comprise the largest volume of equipment processed by IBM's PELM centers, and the product range covers all IBM products from typewriters up to large mainframes. Equipment varies from young (one to three years) to very old (more than 15 years).

Customer-owned equipment comes from product take-back programs offered to IBM customers in 17 countries.[6] Terms of these programs vary by regional market needs and regulations. For example, in Germany, IBM charges the cost for end-of-life management to business customers, which includes the transportation cost. This approach only makes sense for cases where the number of discarded products justifies direct transportation to the dismantling center. For private customers (consumers), IBM offers a solution, which is based on "shared responsibility." The customer delivers the discarded product to the dismantling center and IBM takes care of the end-of-life management without charging for the recycling cost. A Web-based application supports the early notification to the dismantling center and assists the customer in completing the correct shipping papers. In the United States and Canada, the consumer pays a fee that includes the cost of shipping the system to the dismantling center.

In some countries where national recycling schemes already exist to support legislation, IBM normally participates in common recycling schemes—if they provide for the economically best solution. In the absence of these options, business customers can request custom services from IBM to manage the disposition of obsolete equipment. Customer returns are, in many cases, associated with network products and server and storage equipment. Often, large customers who intend to buy new equipment request both delivery of new products and return of old products as a basis for signing a new contract. Customers often want one company to coordinate product disposal from all their global installations. IBM solutions can include product return in many countries with transportation to a PELM center.

Customer-owned returned equipment is generally between three and 15 years old and commensurate with tax regulations for depreciation. Typically, these products are disassembled for the recovery of reusable parts and commodity materials.

IBM's experience in demanufacturing operations has been one of continual effort to assess recycling technologies and vendor capabilities around the world to manage commodity material waste streams from product demanufacturing. Recycling infrastructures with appropriate environmental controls are critical to reducing the potential impact of product disposal, but these recycling industries are not equally accessible for all products. Inspection and qualification of environmentally responsible

---

[6] Countries with IBM product end-of-life service offerings <http://www.ibm.com/ibm/environment/products/prp.shtml> (10 April 2003).

recycling vendors is a necessity which IBM requires through another of its corporate instructions. Likewise, effective remanufacturing operations depend on highly developed broker networks to manage opportunities for remarketing used equipment. The relative proportion of products that are channelled to different end-of-life options relies to a great extent on the capabilities of regional support structures for solid waste management. Opportunities to affect the outcome of product impacts from disposal are greatest with the identification and support of re-use and recycling networks.

IBM has demonstrated success in this approach with its PELM network. Through sharing expertise to improve efficiencies of its demanufacturing operations, a group of IBM's largest PELM centers reduced their collective landfill usage worldwide by more than 60 percent in eight years.

Another benefit of information exchange among PELM centers is the insight that these operations provide relative to the consequences for product design. Feedback to development from demanufacturing operations initiates better design for end-of-life management. From their experience, the centers provide recommendations to product development teams to ensure that product design factors and end-of-life issues adversely affecting re-use and recycling efficiencies are addressed early in the design of new products. Figure 1 shows IBM's view of the feedback loop to optimise materials life cycles.

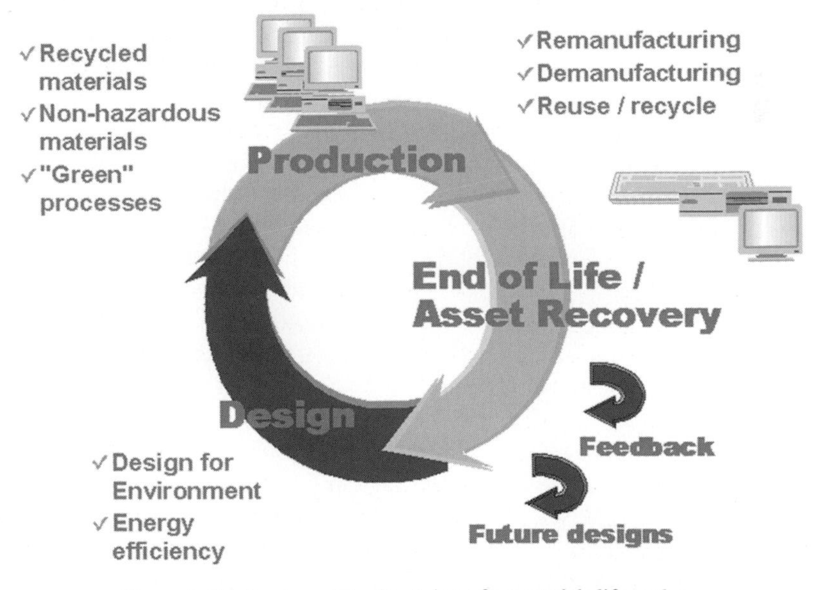

Figure 1: IBM's view of feedback loop for materials life cycles.

9. CONCLUSION

Managing environmental product aspects in IBM includes a wide range of activities, from product design, supply chain management, and manufacturing to sales operations and customer relations management. Product end-of-life management for all products and services is included in these activities. In the view of IBM, the best way to manage product environmental aspects effectively and efficiently is through the integration of the environmental requirements into the product design process and all other relevant business processes. This integration needs to be supported by an environmental management system which has elements for product characteristics. As the business environment changes and business models evolve, there will likely be an increase in the use of original equipment manufacturers (OEMs) and products not directly designed and manufactured by IBM. Based on this fact, a need for harmonization and standardized requirements is seen, and common and global approaches for managing environmental aspects of products will be driven by industry.

Chapter 6

ENVIRONMENTAL MANAGEMENT AT FUJITSU SIEMENS COMPUTERS

Harald Podratzky
Fujitsu Siemens Computers, Germany

Across Europe, the current discussion on environmental protection in the information and telecommunications (IT) industry has largely been limited to the issue of electronic waste. The effect that IT products have on the environment, however, is often underestimated, since it is during a product's entire life cycle that it has an environmental impact, not just at the end of its useful life. It is therefore important for manufacturers, in considering their corporate responsibility towards the environment and society, to be aware of the effects that the company's products have on the environment at every stage of their life cycle.

In the following sections, the practical implementation of environmental protection concepts at Fujitsu Siemens Computers will be described. After a short introduction to the company, the historical development of its environmental protection strategy will be explained and complicated issues concerning the economy and ecology will be discussed.

Furthermore, the company's environmental protection concepts will be presented, and the procedures involved in reducing the environmental impact of product life cycles will be described. The following section will illustrate how environmentally important issues are tied into business processes. This ensures that measures are regularly put into practice and that environmental efficiency is continually improved.

1. A BRIEF INTRODUCTION TO FUJITSU SIEMENS COMPUTERS

Fujitsu Siemens Computers is Europe's leading computer company. It has been jointly owned by Siemens AG and Fujitsu Limited since a merger in October 1999 of Fujitsu Computers Europe (Ltd.) and Siemens Computer Systems, formerly known as

Computers and the Environment: Understanding and Managing Their Impacts
Edited by Ruediger Kuehr and Eric Williams, pages 99–130.

Siemens Nixdorf Informationssysteme AG (founded in 1990 after the acquisition of Nixdorf Computer by Siemens AG).

The services provided by Fujitsu Siemens Computers encompass IT and IT infrastructure solutions for corporations, medium-sized companies, and individual consumers. It is active in IT markets in Europe, Africa, and the Middle East, and profits from the global cooperation and innovative strength of its shareholders, Fujitsu Ltd. and Siemens AG.

Fujitsu Siemens Computers offers one of the most comprehensive product portfolios in the computing business—from Intel and Unix servers through to mainframe computers and data storage solutions. Furthermore, it is one of the leading providers of mobile products, workstations, and personal computers (PCs) for business, as well as being number one in Europe's consumer market (Fujitsu Siemens Computers 2002).

In the company's 2001/2002 fiscal year, it secured a turnover of E.U. €5.5 billion (U.S.$4.6 billion) and employed over 7,000 people. Development and production facilities are based in Germany in Augsburg, Munich, Paderborn, and Sömmerda, while marketing, sales, and service centers are located more regionally.

2. THE HISTORY OF ENVIRONMENTAL PROTECTION AT FUJITSU SIEMENS COMPUTERS

This section outlines the development, evolved from many separate strategies, of a system that encompasses the life cycles of the company's complete product line and all internal operations.

The development of environmental protection strategies within Fujitsu Siemens Computers and its forerunner companies has a long tradition. Early on, measures were introduced within various environmentally sensitive areas to tackle specific problems. These individual measures were merged over time, evolving into a wide-ranging environmental protection system covering the entire production process. By the end of the 1980s, the practise of recycling used and old appliances for resale and re-use was established. From these early experiences with recycling, the importance of issues, such as dismantling and recycling, in the product development process was recognised.

Many of the initial measures to reduce the company's impact on the environment were driven by business considerations. As the public became more aware of the environment, so too did the decision-makers in the company's upper management. As part of society, they took it upon themselves to minimize the environmental pollution caused by processes within their scope and to exceed the standards required by law. In particular, Mr. Hartwig Rogge, director of development and production at the time (at Siemens Nixdorf Informationssysteme AG), by his personal commitment, became an important figure in promoting principles of environmental protection.

When the EMAS[1] Standards for Environmental Management Systems (European Council 1993) and the ISO 14001 international management standard were introduced in the mid-1990s,[2] most of their elements were already included in the company's management system. It was only necessary to collate and unify the separate elements and integrate them into a comprehensive system. The existing perspective of individual responsibility toward the environment and society was written down as an environmental policy.

"We feel responsible for the care of our environment and the prudent use [of] its natural resources...We will endeavour to avoid environmental pollution or to reduce it to a minimum, and in so doing, to exceed the standards required by law." (Siemens Nixdorf Informationssysteme 1996)

From the outset, an important aim accompanying the implementation of environmental management systems was to profit from any economic benefits arising from the systematic introduction of environmental standards. Another deciding factor for establishing environmental management systems was the opportunity to attain a positive, environmentally conscious public image.

In this way, a number of locations of the former Siemens Nixdorf Informationssysteme and Fujitsu GmbH were validated and/or certified according to the EMAS or ISO 14001 standards. In a following development, the synergies of the systems of quality control and work health and safety were harnessed to create an integrated management system. This complicated progression will be explained below, along with a number of examples.

2.1 Typical Occupational and Environmental Protection

The beginning of the 1980s saw the dawn of data processing using large computer systems. The production facility for data processing systems at Siemens AG was located in Augsburg in Werner-von-Siemens-Straße 6, where numerous production stages were carried out, from the manufacture of circuit boards to computer assembly and system tests.

In a corner of one ancillary building was the so-called separation point, where all leftover process components and materials were sent. Some were re-used as spare parts in the repair shops. The remaining parts were broken down into smaller pieces and then sold as scrap metal. When the opportunity arose, some re-useable elements were sold to employees. Any remaining material was then disposed of as waste.

The term *recycling* did not have the familiarity then that it has today, but with the economic rationale of the time, the principle of "re-use before recycling before disposal" was established, which was an improvement for the environment.

Another area of the facility was called the return point. Used appliances were delivered there for refurbishment, where they would be refitted to the latest technical

---

[1] EMAS stands for Eco-Management and Audit Scheme.

[2] ISO stands for International Organization for Standardization.

specifications and sent back into the marketplace. This particular measure was also a specific result of economic considerations, but it had a positive effect on the environment too.

The galvanizing and printed circuit board manufacturing plants were overseen by an official responsible for the safe use of chemicals. When using hazardous materials, the requirements of the health and safety regulations and environmental protection legislation are closely connected. Alongside the responsibility for protecting the health of employees, another important factor was to guarantee that all legal environmental control requirements were being followed. The person who originally carried out this work, Mr. Alois Hampp, eventually became the manager of the environmental protection department at Siemens Nixdorf Informationssysteme AG. Many of the solutions and concepts described in this chapter can be traced back to his initiative and commitment.

A few years later, the production of electronic components was moved to a new location on Bürgermeister-Ullrich-Straße, where a comprehensive analysis of the environmental impacts of the production process was undertaken. Among the issues it revealed was that during the production of electronic components, stencils used for imprinting with solder would occasionally cause a polluting mess. In addition, the solder frames used in wave soldering had to be cleaned on a regular basis.

In the cleaning process, a water-based alkaline cleaning solution was used. The resulting effluent contained not only the cleaning solution but also the rinsed-off impurities. The level of hazardous materials in the effluent was found to be well in excess of the levels permitted for disposal in the regular sewage system. In the old facility, all effluent produced would have been treated in the main sewage treatment plant. In the new facility, which had previously been used for assembly work, a solution had to be found.

The effluent was collected in special tanks, distilled, and then pumped into a circulation system. After the distillation process, only 1 percent of the contaminants remained, and this was disposed of as hazardous waste (Figure 1).

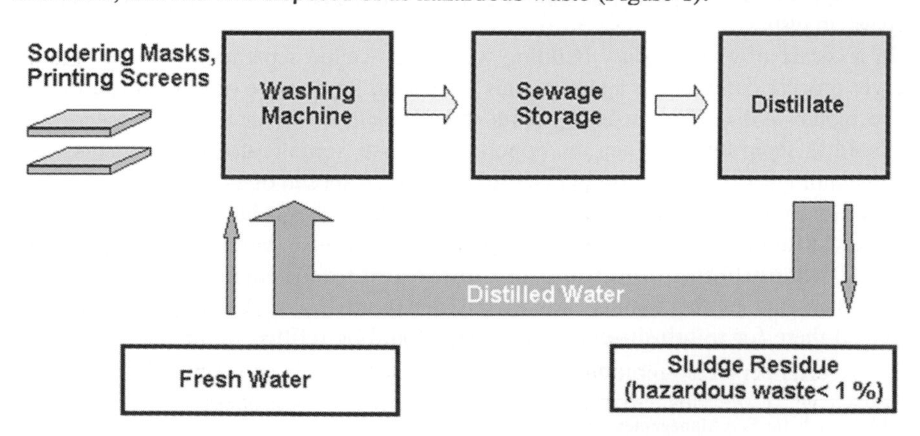

Figure 1: System board production—cleaning without discharging effluent.

2.1.1 Another Example of the Company's Waste Management Practices

The majority of waste produced in the assembly of personal computers and servers largely consists of packing materials that components are delivered in, with a large portion made out of polystyrene. Due to the daily assembly of several thousand computers, the equivalent of a truckload of polystyrene waste is produced every day. Despite its low density, a considerable area is required to collect and store it. The solution to this problem was a machine developed to mechanically break up the packaging material and compress it into small bricks. The increase in density of 1 to 60 reduced the space required in transportation of the waste by the same factor.

Because of its higher density after compression, the polystyrene loses its foam-like structure and cannot be used again for the same purpose. Instead, the material is broken up and granulated to produce polystyrol, a raw material used in various plastic material products. Although this solution produced an environmental benefit by significantly reducing hauling requirements, it has not been globally adopted due to the relatively small amount of materials involved. The Augsburg processing plant remains unique despite having been in operation for over ten years.

2.2 The Beginnings of Resale and Recycling

The Fujitsu Siemens Computers resale and recycling centre on Frankfurter Weg in Paderborn occupies the site of a used-appliances facility set up in 1988 by the Nixdorf computer company.

Formerly, the re-use of computers and components was the top priority. However, when re-use was not possible, the components were more often disposed of than recycled. When Siemens Nixdorf Informationssysteme AG was founded in 1990, a new centre was built to deal with used appliances returned from clients under the "re-use before recycle before disposal" principle.

Along with setting up the company's own recycling centre came the goal to provide customers with the opportunity to return old appliances. Another goal was to obtain spare parts. At the beginning of the 1990s, a discussion began in Germany on the introduction of an electronic waste directive which would oblige manufacturers to take back old appliances free-of-charge, but it was never introduced. At the end of 2002, a similar one, the Waste Electric and Electronic Equipment Directive (WEEE) was endorsed by the European Union (see also Section 6), with the requirement that it must be integrated into German law within 18 months. With this in mind, Siemens Nixdorf pursued the goal of meeting the expected legal requirements through its recycling activities and thereby profit from any economic benefits that could arise.

The continual development of recycling technologies in Paderborn is closely connected to the commitment of the longstanding director of this department, Peter Burgdorf. His personal ambition was not only to achieve his economic objectives, but also to strive for the most environmentally friendly recycling process.

In the first few years, the emphasis of recycling lay in recapturing metals. The computers at this time contained precious metals, including gold, silver, and palladium, and other metals such as copper, aluminum, and steel used in various components. These re-used metals made up only one-third of the computer's total weight. The rest of the material was sent away for disposal.

From 1993, special attention was given to the recycling of the high-quality plastic material used in computer casings. After a number of pilot projects on manually-operated plastic material recognition systems, 1996 saw the introduction of an automated plastic material recognition and sorting facility—the first facility of its kind in the world.

2.3 Environmental Responsibility in Product Development

With increasing experience, the recycling centre operators noticed large differences between the environmentally friendly properties of otherwise similar products. It was thereafter considered necessary to exert an influence on the development of appliances, spawning the idea of an internal company construction standard.

At the beginning of the 1990s there were no standards yet that defined the environmental requirements of computers. The Siemens Nixdorf (SNI) standards for environmentally friendly product design were developed by an in-house committee comprised of not only environmental protection experts, but also professionals in product development, product planning, and recycling. The first version was published in 1992 with the following emphasis:

- Rules concerning easy-to-dismantle design
- Assessment and selection of plastic materials with regard to flame retardant content
- Labelling of plastic parts
- Avoidance of hazardous substances in imported components

Many of the requirements from these original SNI standards were later adopted in today's environmental standards for computers.

An important part of the standard was a list containing prohibited dangerous substances. It was developed with the aim of providing a contractual basis with the suppliers of parts and components and to agree on the exclusion of the listed substances. The first edition was subsequently taken up by the whole Siemens group and today is being continually checked and updated. The idea of a prohibited substances list was also accepted by the European Information and Communication Technology Association. (EICTA 2002)

2.3.1 The World's First Green PC

A result of the company's efforts to lessen the environmental impacts of its products were seen at the Cebit computer trade show and exhibition in 1993, where the world's first "green" PC was unveiled. The drive to develop it came from the environmental commitment of the company's decision-makers at the time and their hope of securing a greater market share by catering to the increasing public interest in environmentally friendly products.

> "Since the end of April, the first green computer is available in Germany—the PCD-4Lsl from Siemens Nixdorf. It is produced using recyclable materials and has an energy consumption of up to 90 percent less than earlier PCs, without any noticeable effect on functionality and processing power."[3]

Important environmental features of the PCD 4Lsl:

- CFC-free production (at the start of 1993, this was not yet standard)
- Only one housing material used
- Synthetic material labelled
- Reduction in the number and weight of components compared with earlier models
- Appliance return guarantee for the customer
- Quicker dismantling through the use of clips rather than screws and welding
- 90 percent recyclability
- Up to 98 percent energy-saving through power management
- Noise reduction through the absence of fans (Siemens Nixdorf Informationssysteme 1994)

The construction of the PCD-4Lsl also set new standards. Table 1 shows a comparison to an earlier model.

Table 1: Comparison in the construction of two models.

Computer	**PCD-2**	**PCD-4Lsl**
Year manufactured	1987	1993
Weight	16 kilograms	6 kilograms
Number of assembly parts	87	29
Assembly time	33 minutes	7 minutes
Dismantling time	18 minutes	4 minutes
Number of cables	13	2

This was the first time that power management, then a familiar function in laptop technology, was used in a desktop PC. Depending on which settings were used, a phased shutdown of the computer's hard disk, disk drives, and monitor was possible. In this mode, only the memory and the keyboard controller were powered, but as soon as the user touched a key, the PCD-4Lsl returned to its normal functioning mode. The PC could also be re-activated from its standby mode by incoming fax or modem signals. The energy consumption of the various running modes is shown in Figure 2.

When the Siemens Nixdorf green PC came on the market, no symbol was invented yet to indicate a product's environmental value. This product had a strong influence on the development of a list of criteria for awarding the environmental symbol, the Blue

[3] *PC Magazine* (Hinnenberg 1993).

Angel, to PCs. It was not surprising, therefore, that one year later, the first Blue Angels were awarded to IT products from Siemens Nixdorf.

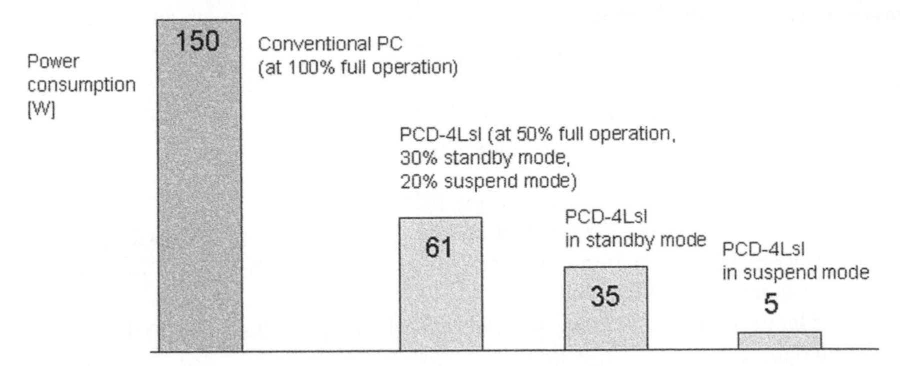

Figure 2: Power consumption of PCD4-Lsl with 15-inch display.

2.4 A Global Approach to Environmental Protection

The previous sections have so far described how, over time, the individual building blocks of an environmental protection framework were created. The logical progression was the establishment of a global approach encompassing the complete computer life cycle "from cradle to grave" (Figure 3).

What follows is a short introduction of the concept (a detailed description of the current situation is given in Section 4).

The concept of an environmentally friendly product design that covers every phase of a product's life is based on the expectations of the customer, and in the widest sense, the whole of society. Another fundamental prerequisite is compliance with all legal standards. To implement such a concept, it is necessary to define the responsibilities and competencies regarding all issues of environmental protection.

The "green" key to success begins in development. In this phase, not only are the quality and economic viability of a product strongly influenced, but also the extent of its future impact on the environment. At this point the material content, manufacturing process, durability, and eventual recycling potential are determined.

In the production process it is important to plan a reduction in consumption of energy and resources, as well as any damaging effects on the environment.

At the end of their practical use, products must be recycled in a way that does not harm the environment. Returning used appliances to the manufacturer (German manufacturers are not yet obliged to do this) creates another advantage. The practical experience gained by the recycling specialists can flow back into the development of newer, more environment-friendly products.

An important role falls to the company employees who, through active commitment, put the processes described above into practice during their daily work.

2.5 Environmental Policy

Next to the active role of every employee, the realization of an environmental protection framework requires the support of management, whose task in realizing this concept is to create a framework of measures from the start and in the ongoing organisation of the company. An important motivating factor for the employees is the role model function provided by the top management and team leaders.

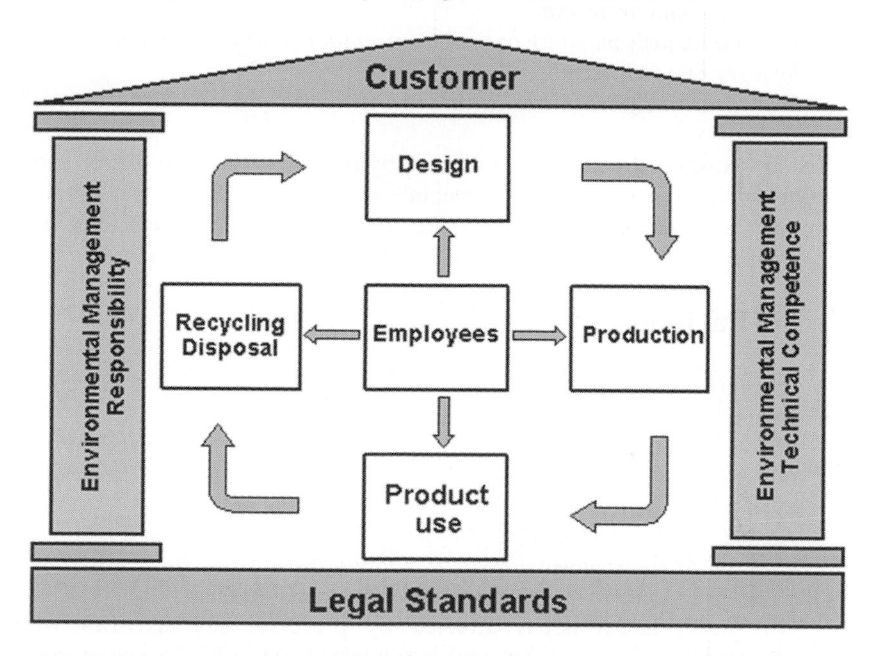

Figure 3: Environmental management system involving the whole life cycle of products.

The first building blocks of the environmental protection concept, laid down since the early 1980s by the forerunner companies of Fujitsu Siemens Computers, were inspired and supported by the personal actions of responsible managers. As already explained in Section 2, this responsibility to the environment and society was later written down as an environmental policy.

With the founding of Fujitsu Siemens Computers came a reworking of the environmental policy. In the environmental management systems of the forerunner companies, the environmental policies were similar in content, but formulated with different words. In addition, the process-specific issues of different company locations were also included in their own company environmental policies.

The goal of the new statement was to create a unified environmental policy for the whole company, which also provided flexibility for the specific aims or particular requirements of individual groups. In this way, the demands of customers, the shareholders (Siemens AG and Fujitsu Ltd.), the general public, and the lawmakers could be met. The environmental policy of Fujitsu Siemens Computers states:

We use our creativity and innovative technological skills to preserve our environment.
Fundamental Principles:
- *We have a responsibility towards society and we commit ourselves to environmental protection.*
- *We continuously minimize our environmental pollution, even more than required by legal regulations.*
- *Environmentally conscious action is an obligation for all members of staff.*

The environmental policy was endorsed by the chief executive officer and formed the basis of the company's environmental management system. Out of the policy, environmental goals were drafted for each company department, and measures to put them in place were issued in the form of environmental programmes.

2.6 Environmental Management Systems

As an introduction, here is a short summary of the significant principles of an environmental management system:
- The assumption of responsibility (voluntarily) to minimize environmental pollution, exceeding the standards set by legal requirements. The conditions are to be laid out in a company-specific environmental policy.
- Analysis and valuation of the company's specific environmental issues.
- Creation of environmental goals to address these issues and the introduction of an environmental management programme to realise them.
- All environmental issues affected by processes and activities within the company will be identified and documented. All affected employees will be involved in this process.
- The maintaining of these self-defined environmental standards will be monitored through internal audits. For emergencies and unexpected work situations, preventative measures should be implemented.
- Through regular management reviews, the efficiency of the whole system can be determined and continual improvement can be encouraged.
- Open communication with the interested public.

(For further information please refer to the ISO 14001 international standard.)

The first preparations for the introduction of an environmental management system began in mid-1995. The EMAS validation of the Augsburg site followed in 1996, with ISO 14001 certification at the end of 1997. The Paderborn recycling centre was also EMAS-validated in mid-1996. The production facility at Sömmerda, originally belonging to Fujitsu, received ISO 14001 certification in 1997.

When Fujitsu Siemens Computers was formed in October 1999, it was decided that all sites would be combined into a unified management system, and the unifying worldwide standard to be adopted for this environmental management system was the ISO 14001 standard.

In the first stage, the development and production sites of Augsburg, Munich, and Paderborn were integrated (February 2001). In the autumn of 2001, the sales and service departments in the five German regions and the production site at Sömmerda were integrated into the system. The result of this process was complete certification in June 2002, under ISO 9001:2000 and ISO 14001, of every Fujitsu Siemens Computers sales and production site. The scope of the certification covers all phases of the product life cycle, including sales, marketing, production, logistics, distribution, and recycling. In addition to every site in Germany, the certification also covers the regional offices in 14 other European countries. More countries will follow in 2003.

The advantages of such a system are the use of synergies in unified processes, the combined application of quality and environmental audits, as well as the opportunity to integrate standards for health and safety at work on a Europe-wide basis. It is presently not possible to certify health and safety standards at work independently.

By certifying a management system covering multiple company sites, the achievement is eligible for a Matrix certification under ISO 9001 and ISO 14001, which is attainable when a company implements a unified management system covering every site and department. By unifying processes and procedures, it is no longer necessary for every site to undergo an annual external audit. Spot checks are an acceptable alternative. Each site is, however, required to have an external audit every three years. The office responsible for monitoring the management system and the Matrix certification receives an external audit annually. This way a pre-arranged check can be carried out on specific documents of the departments not receiving an external audit that year, serving as evidence of a properly functioning management system in these departments.

To apply Matrix certification to an environmental management system, the following conditions are necessary:

1. The management system must be fully integrated into every department. In particular, the following points must be put into practice:
   - documentation of the environmental management system
   - determination of environmental goals
   - environmental management programme
   - every staff member must be trained
   - internal audits must be regularly carried out in all departments
   - management reviews must be regularly carried out
2. In the central office, for every department integrated into the environmental management system, the following documents should be found:
   - department-specific environmental aims and environmental programmes
   - reports of the internal audits and completed improvement measures in each department
   - reports from the management reviews in each department

Under Matrix certification, the possibility of certificate withdrawal from all company sites arises when certification requirements at one single site are not met.

3. THE TENSION BETWEEN ECONOMY AND ECOLOGY

In the reference book, *Environmentally Friendly Product Design* (Quella 1998), one of the authors, Alois Hampp, studied the role of environmental protection as a marketing tool. Up until then, success was determined by the value for money of a product. Even today, environmental protection as an extra buying incentive only becomes attractive when the eco-product has the same value for money as a similar "non-eco-product." An exception to this is when the eco-product offers an additional bonus to the customer, such as reduced power consumption.

Environmental protection is now accepted by today's society as a necessity and no longer called into question; however, this awareness is still seldom integrated into practical action.

In a study by the Marketing Institute at the University of Münster (Quella 1998), the purchasing considerations of German consumers when buying a new PC were investigated. Questions concerning important environmental issues were posed, including recyclability, lower energy consumption, improved environmental compatibility, environment symbols, and a return-for-disposal guarantee. These issues were considered of average importance by business customers and of limited importance by home users.

The development of environmentally friendly high-performance products may create a positive image for the manufacturer, but create no extra turnover. The products are often praised, but not bought to the same extent. The customer is still not prepared to pay more for a product incorporating environmental protection features than for a conventional product.

The first green PC produced in 1993 by Siemens Nixdorf set a milestone in the development of environmentally friendly products and was enthusiastically greeted by environmentalists. The sales figures, however, were lower than expected. Even the more renowned environmental organisations, which enjoyed a high public profile in Germany, still chose cheaper, less environmentally friendly alternatives at this time.

Since then, there had been little change in this general attitude by the end of 2002. Even today, customers who are otherwise environmentally aware are not always willing to pay a little more, for example, for a PC with a lead-free soldered system board.

To ensure the further development of product-related environmental protection, companies are being encouraged "to combine ecology and economics with innovative concepts" (Quella 1998).

To realise this demanding challenge, Fujitsu Siemens Computers employs two approaches:

- a global consideration of the complete product lifecycle, and
- integration of environmental concepts into every internal company process.

These two approaches complement one another in many ways, and the next two sections of this chapter are dedicated to them.

4. A GLOBAL APPROACH TO PRODUCT LIFE CYCLE

The global approach to the product life cycle was briefly discussed in Section 2.4. Current practices at Fujitsu Siemens Computers will now be described.

4.1 Customer Requirements

The most important marketing task, from the environmental protection perspective, is the survey and analysis of customer requirements and wishes and the incorporation of this information into the product development process. Currently for PCs, the Blue Angel (RAL Deutsches Institut 2002) in Germany and the Nordic Swan (SIS Ecolabelling 2002) in the Scandinavian countries have been established. The requirements for receiving the energy-saving label, Energy Star (EPA 2002), which originated in the United States, are included in the requirements of the Blue Angel and Nordic Swan labels and are therefore fulfilled. Another standard to which Fujitsu Siemens Computers products conform to is the TCO'99 (TCO 2002) for monitors. The TCO'99 label is awarded by the Swedish Confederation of Professional Employees and is now established as an international standard.

In addition to the norms and standards described so far is the "eco-declaration" published for every new product. These manufacturer declarations on environmentally important features of their IT products were instigated by the Scandinavian IT Association (IT-Företagen 2002). The aim of creating these eco-declarations was to provide a summary of each product's relevant environmental information for the customer and the general public.

When considering the environmental attributes of products required by European customers and overall attitudes towards this issue, it is possible to determine a north-south dividing line across Europe. Germany's "geographic" position (the subjective view of the author) is in the middle.

Interest in environmental protection is shown mainly by public organisations, insurance companies, and banks, as well as large companies in various sectors. Advertising from these companies often contains questionnaires relating to the Blue Angel, the eco-declaration, or other company-specific interests. This would indicate that the marketing departments of these companies take the issues seriously. Companies certified under ISO 14001 show an active interest in the environmental management of their suppliers. Enquiries are regularly made concerning the return of used appliances. Some public organisations have introduced guidelines stating that the purchase of new equipment should include plans for its disposal at the end of its useful life. In this market sector, conforming to environmental requirements is often a prerequisite for consideration in winning a contract.

Less interest is shown by home users. In this market sector, products are marketed for the performance statistics of their components. Eco-labels, like the Blue Angel, are largely unknown. Only one enquiry has ever been made (in 1998) by a large retail chain concerning the Blue Angel. This was a unique request that was not made on successive product lines.

It is of little use to the environmental protection cause to complain about the lack of interest in several customer groups. It remains an important task for the sales and marketing departments, therefore, to actively promote the environmental benefits of Fujitsu Siemens products and to use this strength as a favourable method of comparison with competing brands.

4.2 Product Development

In recent years the interval between new developments being released onto the computer market has shrunk significantly. This has resulted in a stark shortening of the life cycle of individual products. The period of time that a product with a specific configuration is available on the market is now a matter of a few months. In the PC field, the speed of this development is being pushed by the component manufacturers, in particular, the makers of processors, main memory, and hard disks. For Fujitsu Siemens Computers, as a hardware developer, integrating these components into their products at the right time is essential for survival in the marketplace. These market demands require a development process that is strictly organized and timed to perfection. A delay of only a couple of weeks in relation to the competition's progress can lead to serious economic consequences.

In this dynamic process it is still important to consider the environmental impacts of the products. In the last ten years since the introduction of the in-house development guidelines, a number of key principles have changed, which Fujitsu Siemens Computers took into account in a comprehensive revision of the standard in mid-2002.

The aim of the revision and new presentation of the guidelines was to take advantage of the opportunity to use environmentally responsible product design as a marketing tool, to secure an economic headstart after the introduction of legislative requirements concerning the taking back of used appliances, and to ensure continual improvement by applying the experience gained from the recycling process.

Due to the changed market conditions, the following points are reflections of the supplemented and revised guidelines:

- *A flexible definition of the aim "environmentally responsible product design" for the different product and market segments.*
  Experience has shown that it is not possible to apply a unified environmental standard across an entire product range (from the smallest PC to the largest mainframe computer). Every product group should be assigned its own standards based on its individual environmental impacts. Additionally, the specific requirements of the various market segments need to be determined.
- *Use of pre-defined and market demanded standards.*
  Since the first publication of the design standard in 1992, a number of environmental standards have emerged. Where possible, these standards should be implemented. The use of common standards increases the supplier's

acceptance of environmental requirements, avoids the potential extra costs incurred by special requirements, and makes the work of the buyer easier.

- *Standardized documentation of important environmental characteristics in the form of an eco-declaration.*

Table 2 is a summary of the requirements set out in the guidelines for environmentally responsible product design for the various product groups.

Due to the large numbers of parts and components involved, the emphasis on environmental protection activities has shifted noticeably from production into development and procurement. An important factor is the control over constituent materials used in the procured components. To safeguard this, a list of "to be avoided" and prohibited substances has been produced, which has become an established part of the procurement contracts used by Fujitsu Siemens Computers.

4.2.1 Dismantling-friendly Design

Dismantling-friendly design is a requirement whose practical implementation bring both economic and environmental benefits.

The principle of assembly-friendly design has been familiar for a long time and is often used as a method for optimising costs. An ideal design concept would be where every component would be assembled into the main system unit from the same direction (preferably from above). This way the parts would only need to "drop into place," where they would automatically find their place and be locked into position. This concept forms the basis for the rules of assembly-friendly design; however, it will not be discussed here in detail.

The vision of an ideally recyclable design is represented by the reversed life cycle. The product should easily separated into the minimum number of non-mixed material parts. From this simple consideration, the rules for an environmentally acceptable design were derived.

For the recycling-friendly design of their products, Fujitsu Siemens Computers has taken on fulfilling the requirements of the Blue Angel award (RAL Deutsches Institut 2002). These requirements are listed in Table 3. A distinction has been made here between the "must" requirements and the "should" requirements. The integration of these into the development process will be described in Section 5.

Table 2: Requirements of environmentally friendly products.

	Requirements	Professional PC	Consumer PC	Workstation	Business monitor	Consumer monitor	Business keyboard	Consumer keyboard	Enterprise products	Mobile
1	Hazardous substances: list of prohibited substances, list of substances to be avoided	X	X	X	X	X	X	X	X	X
2	Ecological requirements for packaging	X	X	X	X	X	X	X	X	X
3 3a	"Blauer Engel" (Blue Angel) for personal computers RAL UZ 78	X	(X)		X	(X)	X	(X)		
3b	Blue Angel for portable computers RAL UZ 93									(X)
3c	Server, workstations, and main-frames according to RAL UZ 78 without clause 3.9 noise emission and 3.11 energy consumption			X					X	
4	Nordic eco-labelling (Nordic Swan) for personal computers	X	(X)		X	(X)	X	(X)		(X)
6	Eco-declaration	X	X	X	X	X	X	X	X	X

Table 3: Requirements of the Blue Angel standards.

	Requirement	Parts affected	Must/Should
	A: Build specification and connections		
A1	Are modules made of incompatible materials connected in ways easy to separate or via separation aids?	Housing, chassis, electronic components	M
A.2	Are components that contain hazardous materials easily accessible?	Electronic components	M
A.3	Are detachable connections easy to locate?	Housing, chassis	S
A.4	Can dismantling take place with universal format tools?	Housing, chassis, electronic components	M
A.5	Were the necessary access points and working room provided for tools used in dismantling?	Housing, chassis, electronic components	M
A.6	Are all detachable connecting elements axially accessible?	Housing, chassis, electronic components	S
A.7	Are uniform screw heads (e.g., cross recessed heads) used for the connection of modules?	Housing, chassis, electronic components	S
A.8	Are at least half of the detachable connections between plastic parts plug or snap connectors?	Housing, chassis	S
A.9	Can parts fixed by multiple snap connectors be opened at the same time by one person?	Housing, chassis	S
A.10	Can the supporting surface be retained during the entire dismantling process?	Whole unit	S
A.11	Are electronic components kept away from the housing?	Housing, electronic components	M
A.12	Was a test dismantling carried out (e.g., according to A.1 to A.11) by the manufacturer and documented with particular regard to potential weak points?	Whole unit	M
	B: Choice of material and labeling		
B.1	Is the manufacture of similarly functioning parts being restricted to the use of one material?	Housing, chassis	M
B.2	Are components made of the same material given the same or a similar colour?	Housing, chassis	S
B.3	Is the imprinting of plastic material components reduced to a minimum (e.g., manufacturer's name)?	Housing	S
B.4	Are high quality, recyclable materials and material compounds being used?	Housing, chassis	M
B.5	Is the proportional use of recyclate possible?	Housing, chassis	S
B.6	Are plastic material parts being labelled to ISO 11469 standards?	Components heavier than 25 g	M
B.7	Was the choice of materials (e.g., according to B.1 to B.6) used and presented in written form?	Housing, chassis	M

(cont.)

(Table 3 cont.)

	C: Durability and refurbishment potential		
C.1	Is the product built with a modular design?	Chassis	M
C.2	Can the main system tolerate improvements to the system's performance?	All parts	M
C.3	Is the system compatible for functional expansion?	All parts	M
C.4	Are the refurbished parts planned and permitted in the specification?	Chassis, electronic components	S

In addition to this, further experiences from the recycling process at Fujitsu Siemens Computers were summarised in the "add-on" checklist shown in Table 4.

Table 4: Checklist of additional criteria used at Fujitsu Siemens Computers.

	Checklist
1.	Does the product contain any materials found on the prohibited substances list?
2.	Does the product contain any materials found on the list of substances to avoid?
3.	Has a guarantee from the manufacturer/supplier been requested to conform to the prohibited/avoided substances list?
4.	Is it possible to remove components without damaging them?
5.	Will any signage be placed on the housing material?
6.	Has the requirement for plastic material recycling been punctually registered?
7.	Is it being promoted that original equipment manufacturer parts and appliances can be sent back to the manufacturer under the premise of recycling and environmentally responsible disposal?
8.	Are the current costs of recycling and disposal being investigated? Please fill the space!
9.	Have all the moulded plastic parts got the same color?
10.	Were standard parts and components given preference?
11.	Does the base unit provide a stable foundation?
12.	Was the number of parts minimized?
13.	Was the energy consumption optimized?
14.	Were noise emissions optimized?

4.2.2 Additional Criteria for Environment-friendly Design

The following extra requirements were adopted from the award criteria of the "Blue Angel" for workstation computers RAL UZ78.

4.2.2.1 Modular Design and Refurbishment Capability

- The control unit is of modular design.
- Exchanging modules is possible without using specialist tools.
- The improvement of performance is possible by
  - upgrading to a higher processor performance (optional),
  - upgrading the external CPU cache (optional),

- upgrading the graphic specification (optional),
- enlarging the amount of memory,
- installation, exchange, upgrading, or connection to a server, and
- the existence of at least one free motherboard slot.

- The guarantee on the appliance lasts for three years.
- Once the production of an appliance has ceased, there is a provision for the replacement of parts for at least five years.
- Information concerning durability in the product documents.

4.2.2.2 Reducing the Number of Plastic Materials Used in Housing Parts

- Larger housing parts made from plastic materials to consist of one homopolymer, copolymer, or a polymer blend.
- Housing made of plastic materials are to consist of a maximum of two separable polymers or polymer blends.
- Larger housing parts made from plastic materials are to be designed so that by using existing technology for the creation of high quality, durable plastic products, the material can be used again.

4.2.2.3 Marking of Plastic Materials

- Plastic material parts that weigh more than 25 grams or cover an area of more than 200 square millimeters must be marked in accordance with DIN ISO11469.

4.2.2.4 Material Requirements for the Printed Circuit Boards

- The carrier material of printed circuit boards must not contain any polybrominated biphenyls (PBB), polybrominated diphenyl ethers (PBDE), or chlorinated paraffins.

4.2.2.5 Noise Emission

- The "declared sound intensity-level" (LWAd) may not increase above 48 decibels in idle mode and 55 decibels in operating mode.

4.2.2.6 Batteries

- Complimentary return of all batteries/rechargeables.
- The batteries/rechargeables do not contain cadmium.
- The mercury content must satisfy EU directive 98/101/EEC.
- Batteries/rechargeables containing heavy metals are to be labelled in accordance with directive 93/86/EEC.
- The batteries and rechargeables can be exchanged by the appliance user.
- Non-exchangeable batteries and rechargeables
  - do not contain cadmium, mercury, or lead
  - have a lifespan of at least ten years
  - battery replacement takes place without a change of circuit board

4.2.2.7 Energy Consumption

- The appliance conforms to the requirements shown in Table 5.
- An on/off switch is situated on the front of the appliance.
- The power requirement, measured according to the criteria of the Energy Star Office Equipment programmes of the U.S. EPA, is less than 1 watt in the "off" mode, or less than 5 watts in "off" mode with further functions.
- The appliance can be separated from its power supply for at least four weeks without causing damage.

Table 5: Power requirements of RAL UZ78 workstation computers.

Maximum connection power	Power requirement in energy saving standby mode	Energy saving standby mode (sleep mode)
<200 watts	<30 watts	15–30 minutes
>200 watts	<15% of the maximum connection power	15–60 minutes

4.2.2.8 Packaging

- The plastic materials used in the packaging contain no halogen-based polymers.
- Each plastic material used in the packaging is labelled according to the current standards of the appropriate packaging regulations.

Following the completion of production, an environmental check is made to assess how the requirements are being met. Subsequently, the application for the Blue Angel or Nordic Swan is prepared. An eco-declaration is then created, describing the product's environmental characteristics.

4.2.3 Current Examples of Environmentally Friendly Development

A current example of environmentally friendly development is the Limited Green Edition Mainboard D1337-A, which came on the market at the beginning of 2002.

With the Limited Green Edition, Fujitsu Siemens Computers has shown that total avoidance of hazardous substances in electronic components is now technically feasible. The company has also demonstrated how learning experiences can be gathered even under mass-production conditions.

4.2.3.1 Choice of Materials

For the D1337 model of the Limited Green Edition, the greatest care has been taken to avoid using lead and halogen.

- **Lead reduction** - By using lead-free circuit boards and lead-free soldering processes, the amount of harmful lead has been reduced from 12 grams per system board to 3 grams.
- **Halogen reduction** - A reduction in the use of chlorine and bromine (halogens used as flame retardants) from about 12 percent to less than 0.15 percent is a response to the requirements of the worldwide standard for the monitoring of halogen-free construction materials (JPCA-ES-01-1999). The circuit board is

listed with the Underwriters Laboratories (UL) under the Fuba label, 3104 ML 94V0 (halogen-free material).

4.2.3.2 Environmental Features

Along with many other premium-line system boards, the Green Edition board from Fujitsu Siemens Computers demonstrates the following environmental features:

- **Thermal Management** regulates the fan controls and therefore reduces noise disturbance. Only when the processors have heated up significantly are the fans switched on. By reduced CPU activity, the fans (in silent mode) are completely switched off.
- **HDD Silent Mode** avoids unnecessary noise emissions from the hard disk.
- **IA-PC** (Instantly Available PC) reduces the PC's energy consumption from approximately120 watts in active mode to 5 watts in standby mode.

The Limited Edition Green Board was produced and sold in a batch of 1,000 units. The positive experiences of this persuaded Fujitsu Siemens Computers to integrate a newer model into the Scenic S2 Green PC, introduced in November 2002.

4.3 Production

Production is the typical operating area of environmental protection activities. Because the most significant environmental impacts originate from production facilities, a majority of environmental legislation created focuses on these facilities.

In the field of computer manufacturing, the last few years have seen a clear trend in a continual reduction in the diversity of production processes. At the beginning of the 1980s, the production facility in Augsburg included a mechanical prefabrication and galvanizing plant, a synthetic material spraying plant, and a varnishing and cable fabrication plant. The circuit boards for the electronic components were also produced at the facility. These were put together, soldered, and built into modules. In the last phase of production, the system units were assembled, tested, and then delivered. The delivery boxes for the internal transport of production materials were constructed in the on-site joinery workshop. Even the popular Bavarian delicacy, veal sausage, was specially prepared for the staff in the facility's canteen.

Due to the reduction in the diversity of production, the risk of environmental impact has also been continually reduced. At the end of 2002, the Augsburg production facility was responsible for computer assembly and the production of its own system boards. All other components, including housing, power supplies, extra electronic boards, and main memory, are procured externally. To guarantee the final specifications of custom-ordered computers, the role of logistics has become ever more important, but this development does not mean that company environmental protection is no longer a priority.

By looking at the input/output balance of the facility (Figure 4), one could simplistically observe that computers are manufactured out of modules and components. Extra inputs to the manufacturing process, however, include sources of energy (electricity, natural gas) and water, as well as work and support materials needed. Alongside the finished products, waste, effluent, and emissions also leave the facility.

Figure 4: Production: input–output analysis.

An important role in environmental protection is to reduce the extraneous inputs and outputs intrinsic to the flow of production.

When the foreman of a computer assembly plant is presented with the target of reducing the amount of waste production by 60 percent, that is an impressive goal, but the foreman's ability to influence the situation is limited. The result would be frustration, followed by resignation. Most of the governing factors affecting the amount of waste produced are already established at the start of the production process. The majority of the waste produced comes from the packaging of imported parts and components. Enough potential still remains, however, to reduce the amount of waste produced, the economical use of work and support materials, and the careful separation of waste.

This example shows that in production the most important environmental effects are already determined in the preceding planning phases. On the one hand, development is affected, but so too is production planning. For these reasons, the concerns of environmental protection will be considered at the start of any new production process.

As an example, the procedure requirements for the qualification of a new production process will be described. The company's internal "three-stage plan" procedure controls the following environmentally important issues at each stage:

Stage 1 *Planning Phase*

In the planning phase the environmentally important effects of the proposed facility are analysed.

- Impact on the environment is checked with reference to a special checklist.
- Use of resources including electricity, pressured air, and water is examined.
- Legal requirements: where necessary approval will be applied for.

- Where applicable, an input/output balance will be created for the materials used and the materials listed in relation to their hazard level.
- The licensing of newly introduced hazardous substances will be promoted.
- The requirement for training on environmentally important issues will be determined.
- Where necessary, the environmental protection representative will be involved.

Stage 2 Realization Phase

- Compliance with conditions for environmental legislative approval achieved in facility construction.
- Permission to use hazardous substances granted when the following points are cleared:
  - Nature of delivery of hazardous materials
  - Storage in hazardous materials repository
  - Internal transport within the facility
  - Use of hazardous materials in production
  - Instructions for working with hazardous materials
- Recycling/disposal of waste (disposal license approved)
- Legal issues to do with transportation cleared
- Employee training

Stage 3 Start-up and Production

- Implementation of environmental legislative requirements at the facility
- Organisation of regular checks required
- Implementation of regular surveillance (where necessary)
  - Internal company monitoring
  - Creation of company journals
  - External monitoring
  - Creation of a facility register
  - Miscellaneous conditions on production operations

The reduction in the impact that the production process has on the environment will be illustrated below with some examples.

4.3.1 Practical examples

4.3.1.1 Repeat-use Packaging for Chassis

As already explained, the majority of waste produced in an assembly facility is the packaging that comes with delivered components. Therefore, during regular checks of its main suppliers, Fujitsu Siemens Computers initiated extra checks to assess the environmentally friendly nature of all incoming packaging. Suppliers are rated highly when they use either repeat-use packaging or when they take the packaging away after delivery. For suppliers based in North America or East Asia, however, taking back packaging from Germany would not present the most ideal ecological solution, and this is taken into consideration in supplier assessments.

With the "just-in-time" linkage between the production line and computer housing delivery, which involves repeat-use packaging, the total volume of waste can be reduced by up to 10 percent. Through continued concrete measures to avoid waste production, the volume of waste produced for each computer has been reduced from 5.5 kilograms in 1993 to less than 2 kilograms in 2002.

4.3.1.2 The Usefulness of Critically Analysing Long-standing Processes

The next example shows how the removal of an unnecessary production stage resulted in eliminating the use of specific hazardous materials.

During the preparation of plans for a new assembly line, production planners discovered that a rotating machine had been overlooked and was therefore an unplanned, extra expense. An investigation into the need for the rotation to occur before the packaging process revealed that a particular angle of access was required to optimize the cleaning of the computers after the assembly process. Up to this point, the cleaning was carried out with an alcohol-based solution, which was not always easy to use. There had already been complaints of skin irritation made by employees involved in the process. Other problems included issues of fire protection and disposal, since the solution had to be disposed of as hazardous waste.

The extra investment for the rotating machinery provoked an investigation into the cleaning of the computers. The cleaning process was implemented at a time when the computer housings were stored over long periods, resulting in the accumulation of dust. Now, Fujitsu Siemens Computers receives a delivery of computer housings every two hours under the "just-in-time" delivery programme. The housings, delivered and assembled on the same day, have no time to collect dust. The cleaning process had been continued, however, as no one had become aware that this was now unnecessary to the assembly process. Only with the threat of unexpected extra costs was the issue discovered. The rotating machinery was no longer required, along with the cleaning process, which was then eliminated. This example shows the usefulness of critically analysing even long-standing processes.

4.3.1.3 Reducing the Time Required for Function Testing

The last example in this section shows how an improvement in productivity and quality can have a positive effect on the environment.

Every computer undergoes a function test before leaving the production facility. In the mid-1980s the time required to test a PC was between 24 and 48 hours. The testing area was set up as a fully automated storage facility with data processing connections to a central testing computer. With a continual increase in the number of computers produced, the production engineers and test technicians were forced to reduce the times required for testing. As testing times were reduced even further, however, the concept of a fully automated storage-based testing system was no longer economically viable. The next testing concept they developed and built was a special carriage, where computers were loaded, connected to a power supply, tested, and then sent for packaging and delivery. The carriage testing system only proved to be economically viable when testing times were between two and six hours. But as production figures continued to increase, even this concept was called into question. Currently, the function testing is fully integrated into the production line and testing now takes only 30 minutes.

Through a gradual reduction of the testing time from 48 hours to 30 minutes, the energy consumption required for testing has also been considerably reduced, as has the impact of this process on the environment. This example shows how continual improvement of the efficiency of processes within a company can be of benefit, not only to production figures and productivity, but also to the environment.

4.4 Usability Phase

The usability phase is normally the longest period in the life of a product (excluding single-use products). Before the product reaches the client, however, it will be sealed in protective packaging. The subject of packaging has long been a cause of friction involving various issues, including safety during transport, effect on the environment, and the cost of the packaging itself. Optimizing packaging is an ongoing task for the production technicians at Fujitsu Siemens Computers.

The transport of computers to clients is also an environmental issue that affects the product's impact. The cost of transportation and the environmental impact of the transportation method are, to an extent, influenced by the use of loading space to its maximum potential. The more products delivered per trip, the less the cost and environmental impact per product. This is an important task for the packaging designer, who, through careful measurement, can realize the optimal package stacking capability required for efficient transportation.

During the use phase, customers of Fujitsu Siemens Computers benefit from the environmental characteristics built into their products, including these:

- Reduced energy consumption
- No hazardous substances used
- Lower noise levels
- Reduced radiation pollution
- Power management

These function-related characteristics contribute significantly to customer satisfaction.

4.5 Re-sale and Recycling

A significant component in the Fujitsu Siemens Computers environmental protection concept is the re-sale and recycling centre in Paderborn. Although the return of old appliances is not presently a legal requirement in Germany (end of 2002),

Fujitsu Siemens Computers already offers its customers the opportunity to bring back products at the end of their useful lives to be recycled in an environmentally acceptable way. This return guarantee extends to all products bearing the Fujitsu Siemens Computers logo or logos of the former companies.

The returned appliance process follows a three-stage plan (Figure 5). In the first stage, the appliance is tested to see if, in its existing form or in a refurbished form, it can be sold on the market as a used appliance. If this is not possible, the appliance is taken to a separation centre. Here, the used appliance is tested by a so-called "searching" computer. Since most of the appliances entering the separation centre are company-made products, the computer can recognise whether any of the delivered appliances contain parts that could be used as spares. These parts are then taken out and delivered to the service department for further use. Then comes the third stage of use, where the used appliances are separated into almost 60 predefined material categories.

In handling all these material categories, Fujitsu Siemens Computers works together with a number of external partner companies. Since the responsibility for the proper recycling and disposal of these materials does not end with the handing over of the materials, every partner is subjected to rigorous assessment. The partner must prove that he can deal with the materials in accordance with existing environmental regulations.

Over time, the experience gained by recycling and separating returned products has resulted in a continual increase in the proportion of recycled materials obtained (Figure 6).

The current achievement of nearly 90 percent of materials being recycled cannot be expected from every type of electronic waste. The prerequisite for this high recycling rate is that the appliances are delivered in satisfactory condition, indicating normal use. Through a strategy emphasising the return of one's own used appliances, the opportunities for obtaining spare parts can be used to the full. The most important factor is that the appliances are developed with a recycling-friendly design.

Alongside obtaining spare parts, another advantage offered by an in-house recycling centre is the valuable information gleaned from the dismantling process, which can be used in the development of new products. Recycling specialists are continually involved in all areas of product design, including the formation of guidelines for environmentally friendly product design. As early as the prototype stage, new products are analysed for their recycling-friendly potential.

Design specialists also have the chance to practise the separation of their products on the dismantling line. The practical hands-on tests carried out during training programmes organized by the recycling centre have an important learning effect.

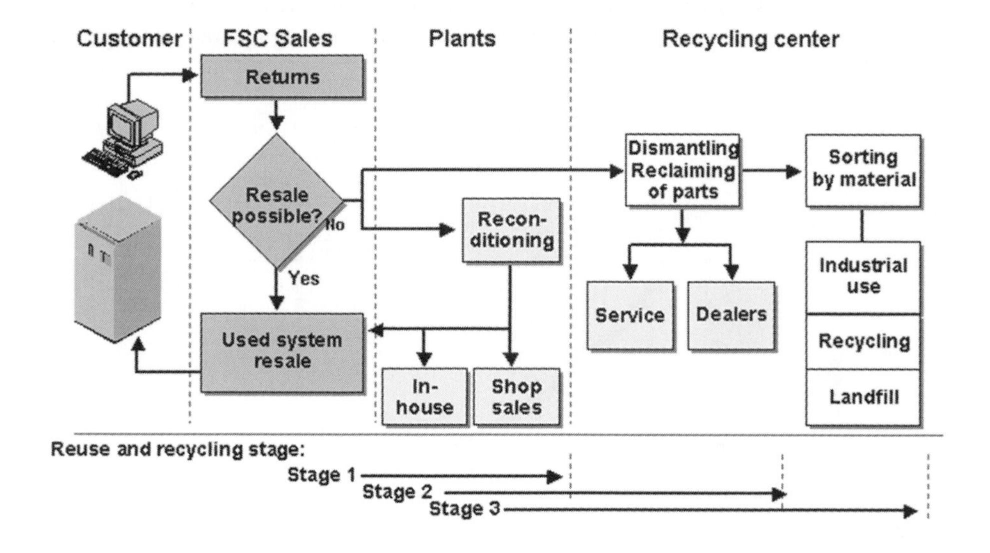

Figure 5: Product-return program in Germany.

5. INTEGRATING ENVIRONMENTAL PROTECTION INTO THE BUSINESS PROCESS

The previous examples show that environmental protection must be integrated into all company activities, processes, and procedures and not restricted to a small, green niche. To avoid duplicating work it is advisable to combine the management systems for both quality and environmental protection.

The long-term success of a company is dependant on how well it controls its processes, and the control of processes is maintained by implementing a management system. The main objective of the management system is to ensure that all processes meet the conditions required of them. This involves the control of product quality and service, as well as environmental compatibility, the safety and health of all employees, the control of commercial risks, and much more. This way the management system can ensure the satisfaction of the customer and the economic success of the company.

Many companies initially considered it sufficient to have their operations certified with the ISO 9001:1994 standard. A reference to processes was indirectly made in ISO 9001:1994, but it was not the focus of the standard. When a management system was introduced under the standard, a very large handbook was often created that was not easy to use, there was not much opportunity to examine processes and procedures, and the main goal of improving efficiency was lost in all the details.

These events resulted in the overall impression that the norm, ISO 9001, was not vital and that some thought of it as a necessary evil or just a bureaucratic formality.

In the mid-1990s, when the standards for environmental management systems were created, they were initially introduced alongside the existing quality management system.

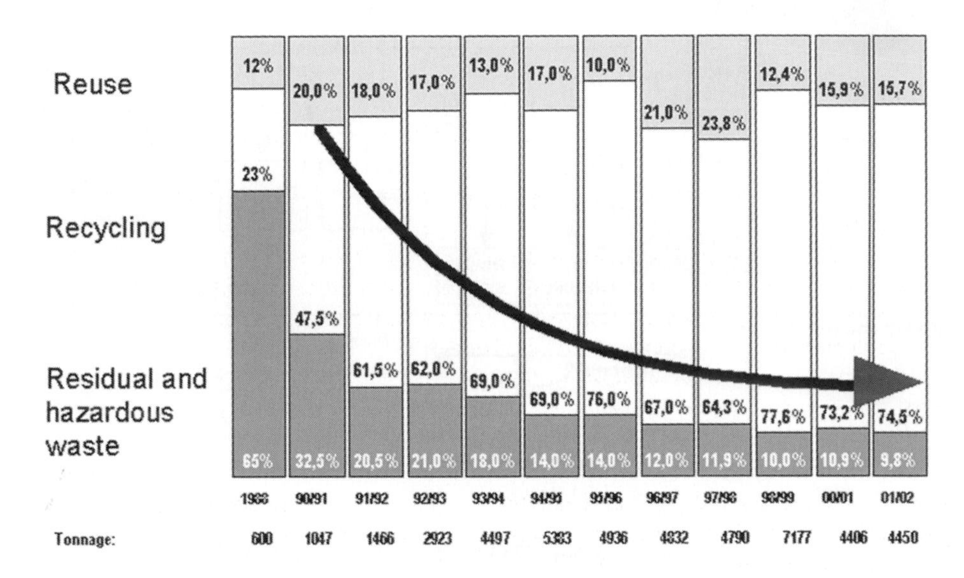

Figure 6: Reduction in the volume of waste in the recycling center.

The fact that the standards were completely different in their structure made it particularly difficult to unify them into one management system. Another reason for two separate management systems was that the ISO 9001 standard was mostly linked to the quality control department, while the environmental protection specialists were normally assigned to the production managers.

Only in the last few years has the EFQM model for business excellence reinforced the emphasis on process theories in the area of quality and the environment.[4] This model eventually led to the re-working of the ISO 9001:1994 standard.

The forerunner companies of today's Fujitsu Siemens Computers quickly recognised that by implementing these norms, they could take advantage of the potential for increasing productivity and quality. Early on, Siemens Nixdorf had focused their management system on the structuring and continual improvement of their processes. When Fujitsu Siemens Computers was founded, these procedures were unified for all company departments and facility locations.

When the ISO 9001 standard was revised in the year 2000, process concepts were given strong emphasis. Due to their existing and actively practised process orientation,

---

[4] European Foundation for Quality Management (EFQM 2002).

Fujitsu Siemens Computers was able to become one of the first companies to be certified under the newly revised ISO 9001:2000 standard at the beginning of 2001.

In the Fujitsu Siemens Computers management system, the issue of environmental protection is properly integrated into the process field. The environmental protection process is a supporting process, with the responsibility to ensure that all other processes are performed in an environmentally friendly way.

The following is a brief description of the Fujitsu Siemens Computers process field. Subsequently, with reference to a few examples, the integration of environmental issues into the processes will be described. The process architecture is structured into a number of layers. (Figure 7 describes the top-level processes.)

- Top level - Describes the complete overview of the central and supporting processes.
- Level 1 - Detailed description of all processes in level 1.
- Level 2 and above - Description of detailed minor processes with information on input, output, activities, responsibilities, and assessment criteria. All important information and documents are described.

The central processes—*concept to product, opportunity to order/delivery to cash, order to delivery,* and *problem to resolution*—begin with the requirements of the customer and end with performance fulfillment for the customer.

The processes of marketing, *global sourcing*, *resource management*, and *business management* are there to provide continuous support to the functioning of the central processes. The environmental protection process is part of the supporting process, *business management.*

The following is a description, using two examples, of how individual environmental issues are integrated into processes.

5.1 Process: Idea to Product

In the Idea to Product process the following points are included in the process description.

Product Definition

Environmentally important requirements are included in the product agreement.

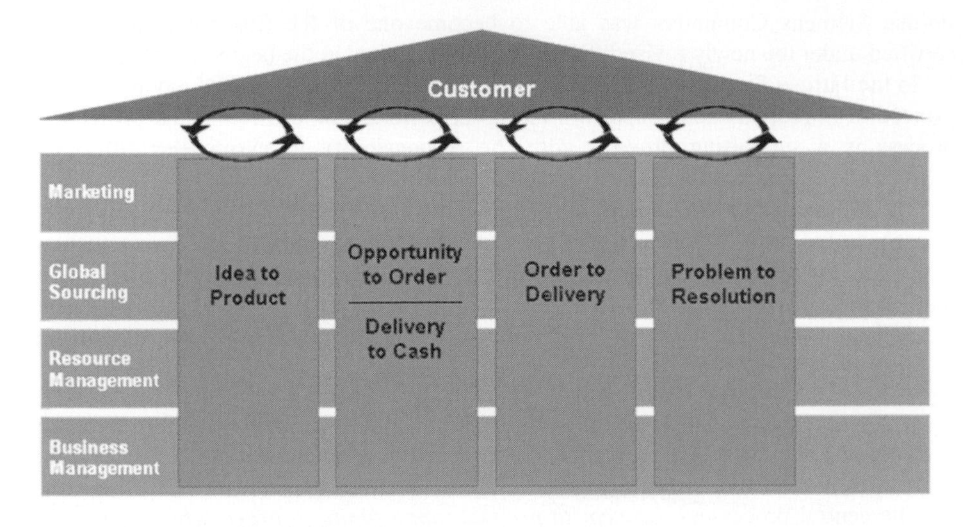

Figure 7: Top-level process architecture.

Check-points During the Development Process

- Supplier confirmation for parts and components concerning the following:
  - Abiding by the prohibited substance and substances-to-avoid list
  - Supplier confirmation regarding plastic materials housing
  - Supplier confirmation regarding printed circuit boards
- Check-lists for environmentally friendly product design
- Recycling analysis for new product ranges
  - Evaluation of recycling-friendly characteristics
  - Calculation of recycling costs

Environmental Requirements

- - Creation of an eco-declaration

5.2 Process: Order to Delivery

The production process is a part of the process Order to Delivery. Within these processes is, among other things, the integration of hazardous substance management.

Within the internal guidelines, the complete passage of hazardous substances through the company is regulated. The following sections are therefore considered:

- Legal requirements
- Definition of responsibilities and competencies
- Introduction of new hazardous substances
- Company processes:
  - Ordering

- Delivery
- Receipt of goods
- Into storage
- Safety precautions during storage
- Surveillance / Controls
- Out of storage
- Internal company transport
- Dealing with hazardous substances in the workplace
- Delivery of old materials (waste) to the hazardous substances depot
- Preparing waste for transport
- Disposal of hazardous waste

6. OUTLOOK

In a field where new and more powerful products are introduced on the market every month, an outlook on the future is somewhat risky. One factor which will have a profound effect is the European WEEE Directive, endorsed during the editing of this book. The directive describes the responsibility of manufacturers in accepting used appliances from consumers at no extra cost. It comes into force in the second half of 2005. Alongside the WEEE Directive, a second one, the Restriction on Hazardous Substances (RoHS) will be implemented. This will state that, from 2006, the use of certain hazardous substances in electrical products, including cadmium, mercury, lead, hexavalent chrome, PBB (polybrominated biphenyls), and PBDE (polybrominated diphenyl ethers) will be prohibited.

In preparation for the WEEE Directive, Fujitsu Siemens Computers has established an internal working group to comply with the requirements.

A further issue that requires extensive development is that of lead-free soldering. To what extent the alternatives to lead in solder will appear on the market before new legislative requirements close in, only time will tell.

A term often used in connection with environmental protection is sustainable development, but it is more than just a fashionable slogan. In a quickly developing global economy, issues of environmental protection need to be considered next to economic and social factors to ensure that future generations also have a share in our world's resources.

Through the distribution of a product's wealth supply chain across the globe, the responsibility of an end-product manufacturer continues to grow. It is no longer enough to ensure that one's own production facilities are clean and that the products are free of hazardous substances. Sustainability involves a responsibility for the production methods and their social side-effects in the countries where components originate. Corporate environmental management must develop in this direction.

In the near future, the tasks awaiting the environmental protection coordinators at Fujitsu Siemens Computers will remain both interesting and exciting.

REFERENCES

European Council. 1993. Council Regulation (EEC) No. 1836/93, 29 June 1993, allowing voluntary participation by companies in the industrial sector in a community eco-management and audit scheme (EMAS).

European Foundation for Quality Management (EFQM). 2002. EFQM excellence model. Brussels: EFQM. <http://www.efqm.org/model_awards/model/excellence_model.htm> (28 December 2002).

European Information and Communication Technology Association (EICTA). 2002. Excerpt of restrictions on substances from legal provisions for special applications in electric and electronic products, update 1 August 2002. Brussels: EICTA. <http://www.eicta.org/content/default.asp?PageID=211> (23 May 2003).

Fujitsu Siemens Computers. 2002. <http://www.fujitsu-siemens.com> (30 December 2002).

Hinnenberg, Ralf. 1993. Stromsparender computer. *PC Magazin* 24, 9 June1993.

IT-Företagen (Association of the Swedish IT and Telecom Industry). 2002. Eco-declaration guideline, version June 2000. <http://www.itforetagen.se/pdf/Guideline.pdf> (28 December 2002).

Quella, Ferdinand. 1998. *Environmentally friendly product design* (in German). München: Publicis MCD.

RAL Deutsches Institut für Gütesicherung und Kennzeichnung e.V (German Institute for Labelling and the Safety of Goods Association). 2002. *Environmentally Acceptable Workstation Computer RAL-UZ78 Edition, February 2001.* <http://www.blauer-engel.de/englisch/navigation/body_blauer_engel.htm> (28 December 2002).

Siemens Nixdorf Informationssysteme (SNI). 1994. *Umweltbericht 1993.* Paderborn: SNI.

———. 1996. *Umwelterklärung.* Augsburg: SNI.

SIS Ecolabelling. 2002. Nordic Swan ecolabelling of personal computers, criteria document version 3.1. Stockholm: SIS Ecolabelling. <http://www.svanen.nu/DocEng/048e.pdf> (28 December 2002).

TCO (The Swedish Confederation of Professional Employees). 2002. TCO'99 ecology for displays, portable computers, system units and keyboards, report no. 5, edition 2, 20 June 2002. Stockholm: TCO. <http://www.tcodevelopment.com/i/arbetsmiljo/index_db.html?tco99.html~ main> (30 December 2002).

U.S. Environmental Protection Agency (EPA). 2002. ENERGY STAR labeled computers and monitor. <http://www.epa.gov/nrgystar/purchasing/6a_c&m.html#specs_cm> (29 December 2002).

Chapter 7

ENERGY CONSUMPTION AND PERSONAL COMPUTERS

Danielle Cole
University of New South Wales, Australia

1. INTRODUCTION

The annual energy consumed by office and telecommunication equipment in the United States during 2000 was estimated to be 97 terrawatt-hours (TWh) per year (1 TWh is equivalent to 1 billion kilowatt-hours), equivalent to approximately 3 percent of the U.S.'s energy consumption (Roth et al. 2002). Personal computers (PCs) accounted for approximately 43 percent of the energy consumed by office equipment—amounting to 41.8 TWh annually (Figure 1).

In the realm of reducing energy consumption, the use of power management functions in office equipment in the U.S. saved 23 TWh per year in 2000; a complete saturation and proper functioning of power management would achieve additional savings of 17 TWh per year. Furthermore, complete saturation of night shutdown for equipment not required to be left on at night would reduce power use by an additional 7 TWh per year. A large proportion of these savings could be accomplished by addressing power management in computers and monitors, because it is estimated that only 25 percent of computers are correctly power-managed, and because power management of monitors reduces energy consumption significantly (Kawamoto et al. 2000).

More than 70 percent of the energy used by office equipment in the United States is consumed by the commercial sector. The industrial and residential sectors make up approximately 13 and 12 percent, respectively, while energy used for network equipment (e.g., routers, switches, access devices, hubs) accounts for less than 5 percent (Kawamoto et al. 2000). These figures vary between countries. For example, Mungwititkul and Mohanty (1997) reported that 90 percent of office equipment in Thailand is currently used in the commercial sector.

Computers and the Environment: Understanding and Managing Their Impacts
Edited by Ruediger Kuehr and Eric Williams, pages 131–159.

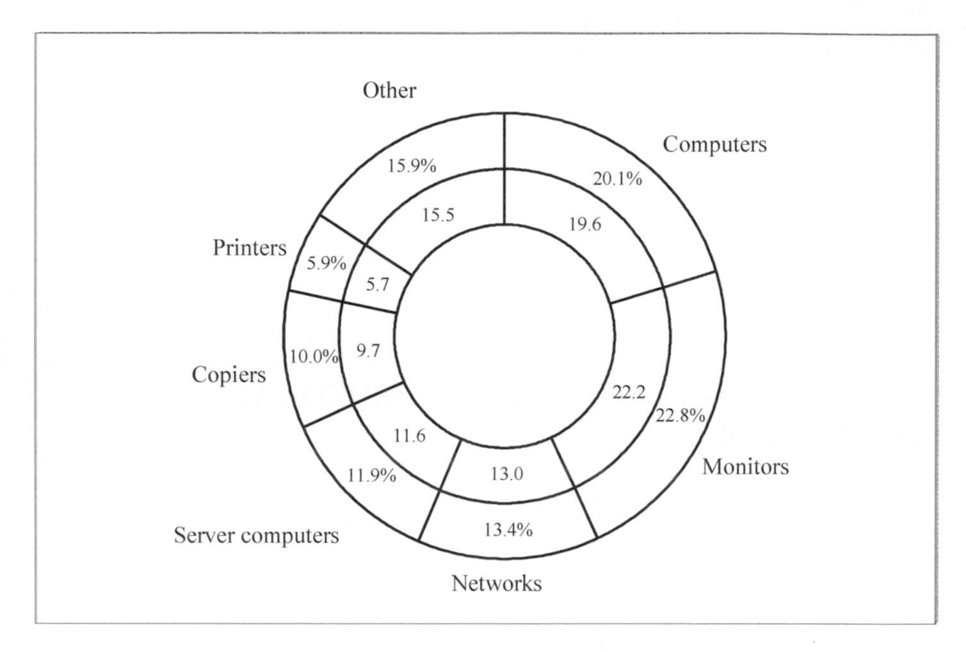

Figure 1: Annual energy consumption of office and telecommunications equipment in the U.S. by equipment type.
Note: The inner ring represents annual energy consumption in TWh, while the outer ring represents percentage of all U.S. office and telecommunications equipment energy consumption.
Source: Adapted from Roth et al. (2002).

Norford et al. (1988) reported that after the 1973 energy crisis virtually all energy use studies and energy consumption programs placed office equipment in the unexamined, residual category of "miscellaneous" or "other." Energy consumption by office equipment was the fastest growing end-user of electricity, of which a significant proportion could be attributed to computers. Norford et al. estimated that electronic office equipment consumed 25 TWh in 1988—one quarter the annual energy requirements of office equipment in the United States today.

In the early 1990s, energy used by office equipment, especially PCs, continued to increase. The rise of the information-based economy, including the Internet, was a major contributing factor, although some authors have overestimated related energy consumption. In 1999, a widely publicized report by Mark Mills for the Greening Earth Society (Mills 1999; summarized in Huber 1999) estimated that the Internet accounted for 8 percent of U.S. energy use, and projected it to rise by 30 to 50 percent in 20 years. The bulk of experts who have looked at this work, however, believe it is largely flawed. A review by Koomey (2000) found that the analysis overestimated the energy intensity of the Internet by a factor of eight, meaning that Internet use currently amounts to little more than one percent of U.S. electricity demand. Further rebuttal of

Mills' claims was presented by Romm et al. (1999). Their research found that in the four years prior to the Internet boom (1992–1996), growth in the U.S. gross domestic product (GDP) averaged 3.2 percent a year, while electricity demand grew 3 percent per year. In the Internet era (1996–2000) GDP growth has averaged over 4 percent a year, while energy demand is growing only 2.5 percent per year. Romm et al. speculated that GDP increased faster than energy demand due to gains in productivity, the replacement of old, energy-intensive computers with more efficient models, and the increasing ability of people to work at home, which reduces both transport and office energy costs (Perry 2001). Laitner et al. (2000) support this argument. They say the decline in energy intensity from 1996 to 2000 was significantly greater than the 2.7 percent rate of decline during the "oil crisis" years of 1973 to 1986. While some analysts attributed the majority of the change to weather, a separate analysis by Laitner et al. indicated that increased climatic temperatures accounted for only 25 percent of the change in the nation's energy intensity. Laitner et al. believed that the remaining energy intensity shift was largely due to the rapidly growing influence of the information-based economy, including the Internet. Contrary to Mills's claims Romm (2002) said that "It [the Internet] appears instead likely to be driving efficiencies throughout the economy that have resulted in the biggest declines in electricity intensity and energy intensity the nation has seen in decades".

Growth in energy requirements for office equipment is beginning to slow as the market becomes saturated (especially for computers and monitors) and as equipment becomes more energy-efficient (Koomey et al. 1995). Nevertheless, the U.S. Department of Energy predicts that energy use by PCs in the commercial sector is expected to grow 2.9 percent per year on average between 1998 and 2020 (Webber et al. 2002). Energy management tools are now standard features in most PCs and are extremely effective in reducing PC energy consumption when idle. There is still potential for additional energy savings, however, via correct use of these power management features and by switching equipment off when not in use.

In this chapter we examine the proliferation of the PC, beginning in the late 1970s. We examine how the PC has replaced typewriters, minicomputers, and mainframes, and look at the emergence of more energy-efficient models of computers such as laptops. In the next section we examine how the power consumption of computers, monitors, laptops, minicomputers, and mainframes has evolved over time. Finally, we look at methods for reducing energy consumption, including energy management options such as the Energy Star program. We also consider the effects of turning a PC off when not in use, as this has the greatest potential for energy savings. Whether or not this action reduces equipment life, however, has generated significant debate.

The majority of information presented in this chapter is sourced from the United States, where most research has been undertaken. The U.S. is also leading the way in PC energy-efficiency through schemes such as the Energy Star program of the U.S. Environmental Protection Agency (EPA). Emphasis is placed upon PCs in the commercial sector, where the majority of energy is consumed by office equipment, especially PCs.

Throughout this chapter, "computer" refers to the processor and related hardware. "Monitor" refers to the visual display unit, which is usually a cathode ray tube. "PC" refers to the entire system—that is, the computer and monitor. These are the official definitions of the terms according to Webopedia (2002); however, "computer" and "PC" are often incorrectly used in the literature to refer to both the computer and monitor or the computer alone.

2. BRING ON THE PC!

PCs first appeared in the late 1970s. The first and most popular PC of its time was the Apple II, introduced in 1977 by Apple Computer. In 1981 IBM entered with its first PC called the IBM PC. There was a proliferation of PCs in the workplace in the early 1980s as PCs replaced electronic typewriters in addition to performing new office functions. The total stock of single-user PCs in the workplace increased more than tenfold, from 1.2 million to 13 million between 1981 and 1986. In 1986, 3.6 million PCs were sold (an estimated 22 percent for replacements and 78 percent for new or expanded use) compared with 1.2 million electronic office typewriters (Norford et al. 1988). There was also a tendency for PCs to displace mainframes in many new and established markets. The number of PCs sold in 1986 was 2.5 times greater than in 1982. Over the same time the number of mainframe units sold grew only slightly from 2,100 to 2,400. PCs were capable of performing many functions of a mainframe at a fraction of the cost. Norford et al. (1988) also reported that some business and technical users had both a PC and shared access to a minicomputer or mainframe; the PC was used either for separate applications or to log-in to a larger machine.

The demand for PCs throughout the 1980s and 1990s was driven by the evolution of the PC, from command line-driven machines with floppy disk drives and capable of limited tasks, to user-friendly, powerful PCs with Pentium processors and add-ons capable of doing anything, from assisting with music composition to editing videos. The decreasing cost of PCs also allowed more people access to PCs and added to their increasing popularity.

During the 1990s, PCs continued to increase in popularity (especially in the residential sector), spurred on by the emergence of the information-based economy, including the Internet. The number of Web sites increased from zero in 1990 to ten million by 2000 (Kawamoto et al. 2000). Romm et al. (1999) estimated that the number of Internet users in the United States soared from 5 million in 1993 to 62 million in 1997, to over 100 million by mid-1999. Kawamoto et al. (2000) estimated that annual shipments of computers increased by a factor of five in the 1990s. Total stocks of computer equipment in 1999 compared with 1981 and 1986 are indicated in Table 1.

Laptops, which are much more energy-efficient than conventional desktop PCs, are replacing the latter with increasing frequency. While Kawamoto et al. (2000) reported that in 1999 approximately one in six PCs in the United States were laptops (Table 1),

Matthews (2001) estimated that in 1998 about one in four PCs sold were laptops (12.3 million laptops compared to 36 million conventional desktop PCs). The Japan Electronics Industry Development Association (1999) reported that laptops made up nearly 50 percent of PC shipments in Japan in 1998 (domestic shipment was 3.566 million units of laptops and 3.972 million desktops).

Table 1: U.S. stocks of computer equipment (thousands).

Equipment type	**1981**[1]	**1986**[1]		**1999**[2]			
	Commercial	Total	Commercial	Total	Residential	Commercial	Industrial
PCs	1200	32,500	13,000	109,110	54,530	47,760	6,820
Laptops	-	n/a	n/a	22,150	16,090	5,300	760
Minicomputers	n/a	375	375	2,020	0	1,520	500
Mainframes	n/a	2.4	24	107	0	96	11

Source 1: Norford et al. (1988). n/a = not available
Source 2: Kawamoto et al. (2000). dash = does not apply

Conventional cathode ray tube (CRT) monitors are beginning to be replaced by liquid crystal displays (LCDs) and, more recently, thin cathode ray tubes (thin CRTs). The advantages of LCDs are their compact size, low mass, negligible electromagnetic emissions, and low energy consumption (requiring 60 to 70 percent less energy than CRTs). They can also be powered up and down more rapidly than CRTs (Nordman et al. 1997). In recent years, manufacturers have made significant improvements in LCDs. At the same time, prices for such displays are falling due to improvements in production processes and an increase in manufacturing capacity, making LCDs increasingly competitive in the market for desktop monitors. In 1993, coloured LCDs had a one to five ratio of cost disadvantage over colour CRTs (Norford and Dandridge 1993). Over the 1995 calendar year, prices for 10.4-inch active matrix screens dropped from over U.S.$1,000 to about $350 (Koomey et al. 1995). Data collected by Roberson et al. (2002) suggests that LCDs are currently twice the price of CRTs. Until the late 1990s, LCDs were only popular in Japan, where 1999 demand was estimated to be 3.4 times 1998 demand levels. Now LCDs are appearing widely in the United States (OEP 1999).

Thin clients, consisting of a screen display, keyboard, and mouse connected to a terminal server that executes applications, look promising for PC energy-efficiency in the commercial sector. They consume 30 to 60 percent less power than traditional PCs and have the added benefits of improved security, centralized administration, and reduced desktop maintenance costs (Greenbergy et al. 2001).

3. HISTORICAL TRENDS IN PC ENERGY CONSUMPTION

Power requirements and the periods of time that equipment is turned on determine its energy consumption. The average annual energy consumption of computer equipment is presented in Table 2. These estimates are based on equipment in the

United States in 1999. Average operating patterns and energy reductions from power management are taken into consideration.

Table 2: Best estimates of unit energy consumption (UEC) for computer equipment in 1999 (United States).

Equipment type	Residential UEC (kWh/year)	Commercial/Industrial UEC (kWh/year)
Desktop computer	49	213
Monitor	57	205
Laptop computer	8.6	24.6
Minicomputer	n/a	5,840
Mainframe	n/a	58,400

Source: Kawamoto et al. (2000).

In the early 1990s, PC energy consumption first entered the literature of the energy conservation communities (Norford and Dandridge 1993). Several office energy-efficiency programs were introduced in the United States and Europe, such as the U.S. EPA Energy Star program (see Energy Star section below for further information). This pushed manufacturers to produce machines with power management features that reduce energy consumption during idle periods. The Energy Star program also influenced manufacturers to produce machines that are more energy-efficient in active mode.

The power requirements of computer equipment have changed considerably since the 1980s, and are discussed in the section below. Power consumption is indicated in two modes. Active mode is when the device is in operation. Standby mode refers to a mode which attempts to conserve power with instant recovery. Lower power modes than mentioned here may also be possible, especially for machines manufactured after 1993 (see section below for further details on automatic power management). Estimates of power consumption are calculated from actual measurements, rather than nameplates. It is well known that the nameplate overstates the actual power consumption of PC systems by a factor of two to four (Norford et al. 1988; Newsham and Tiller 1994; and Wilkins and Hosni 2000). Figures displaying power consumption must be read with caution as data has been compiled from multiple sources and therefore collection methodologies and equipment models may differ. The figure captions provide information about each data source and should be taken into consideration when examining trends.

Later in this section we examine historical trends in PC operating patterns and the additional energy requirements of air-conditioning systems to offset heat produced by computer equipment.

3.1 Computer Power Consumption

The very early computers were extremely inefficient, with low computing power and high energy consumption. The energy efficiency of computers increased from the mid-1980s until the mid-1990s, as demonstrated by time series data sourced by

Koomey et al. (1995) and depicted in Figure 2. During the mid- to late 1990s computer energy efficiency leveled off, with some variation between different models (Figure 2). Average power requirements of computers decreased by almost 50 percent, from nearly 100 watts (W) in the mid-1980s to 50 W in the mid- to late 1990s, and standby power consumption stayed relatively constant at approximately 25 W since its introduction for computers in the mid-1990s. Computer power consumption is on the increase, however, while standby power consumption is decreasing. The new Pentium 4 consumes more power than its predecessors at 67 W in active mode, while consuming only 3 W in standby (Roberson et al. 2002).

In 1988 Norford et al. (1988) reported that newer models of computers with equivalent performance were often more energy-efficient. Initially, these energy-efficiency improvements were driven by the economics of chip manufacturing, as well as by the manufacturer's desire to fit more peripherals into smaller spaces (an effort that required heat reduction and hence efficiency improvements) (Koomey et al. 1995). Later energy conservation awareness programs, such as Energy Star, significantly influenced manufacturers to produce more energy-efficient machines.

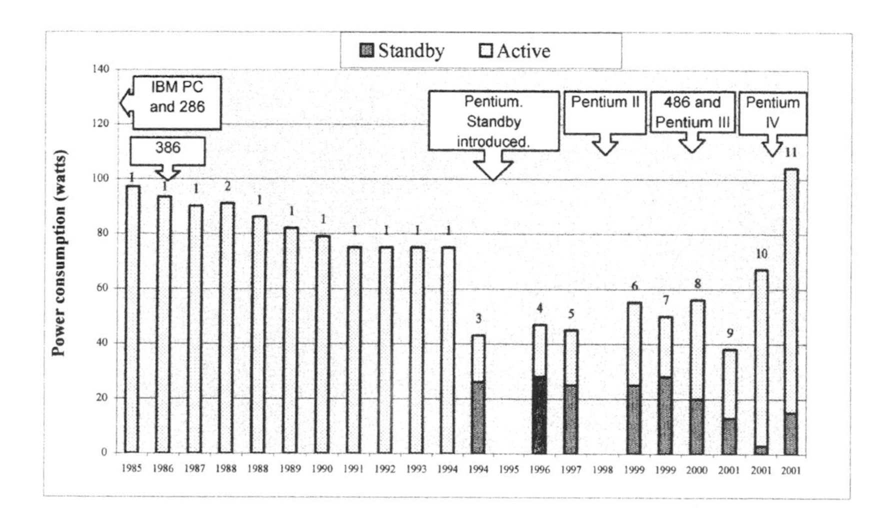

Figure 2: Power consumption of a desktop computer (without monitor) from different studies, according to year.

Note: Text boxes indicate when different processors and standby for computers were introduced. Note that different models and brands have been measured using different methodologies, so the data should be interpreted with caution.

Sources: The numbers above the bars refer to the source of the data.

1 Figures based on measured data by various authors, trade press assessments, personal communications from industry participants (Koomey et al. 1995).
2 Average of 13 computers (Norford et al. 1988).
3 Based on 5 Pentiums (UNSW 2000b).
4 Average of 3 computers (Nordman et al. 1996).

5 Pentium (Nordman et al. 1997).
6 Pentium. Figures based on measurements by various authors (Nordman 1999).
7 Macintosh. Figures based on measurements by various authors (Nordman 1999).
8 Based on average of 20 486 computers (Wilkins and Hosni 2000).
9 Average of 4 Pentium 3s (Roberson et al. 2002).
10 Average of 6 Pentium 4s (Roberson et al. 2002).
11 Average of 4 AMD Athlons (Roberson et al. 2002).

While computers became more energy-efficient in terms of providing more computing and storage per watt of power, however, the computing power also increased. Until recently, advancing technologies have curbed overall increasing power levels so that in the mid- to late 1990s, computers have operated at nearly constant power levels with more powerful microprocessors, more memory, and more disk storage. There is much more variation in the power requirements of the modern computer, due to the addition of consumer-specified features, such as hard drive capacity and add-on cards, which vary power requirements of similar models (Howarth et al. 2000, Nordman et al. 1996). Of the modern computer, Nordman (1999) goes so far as to say that "The variation at any time among models sold seems to be considerably greater than the variation across time."

Reduced computer energy consumption has been achieved through the design of computers with components which require less power (providing it does not take longer for the component to perform the required task). Figure 3 compares power requirements of the components of an average computer in 1988 (as documented by Norford et al. 1988) with those in 2001 (as recorded by Peppercorn 2001). The modern hard drive requires significantly less power than earlier models (10 W compared with 35 W), as does the motherboard (25 W compared with 52 W in 1988). The power requirements of the floppy drive remain unchanged.

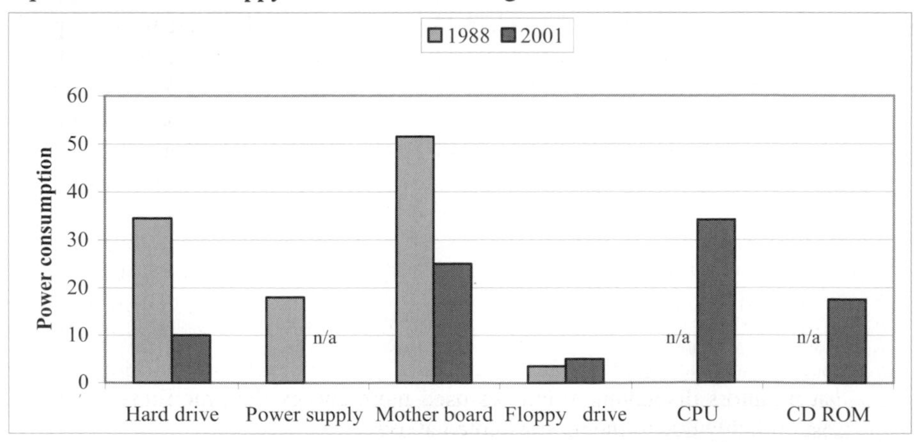

Figure 3: Measured component power requirements for average computer in 1988 compared with 2001. *Source*: 1988 data by Norford et al. (1988), 2001 data by Peppercorn (2001).

The modern central processing unit (CPU) is one of the few computer components that use more energy than earlier models. The CPU was not recorded in the Norford et al. (1988) study because it consumed minimum power—compared with an average of 34 W in 2001. The power required by the CPU can be expressed as "the product of the processing speed, the number of transistors being switched, and the energy required to switch each transistor, which in turn varies with the capacitance and the square of the voltage" (Norford and Dandridge 1993). Capacitance scales with transistor size. The number of transistors in a CPU has increased faster than the transistor size has decreased, as indicated by the growing overall dimensions of CPUs (Norford and Dandridge 1993).

Power consumption of an electronic chip depends mainly on the type of transistor used. Laptop computers rely on "complimentary metal-oxide semiconductor" (CMOS) technology. Prior to the 1990s, most desktop computers used "n-channel metal-oxide semiconductors" (NMOS). Both types of transistors drew similar amounts of power when switched, but CMOS requires almost no power when in a quiescent state (Norford et al. 1988). Use of CMOS transistors in desktop computers has reduced power consumption by around 30 to 40 percent.

3.2 Monitor Power Consumption

While computers have tended towards greater energy efficiency until recently, monitors have become more energy-intensive (Figure 4). In the 1980s, the average monitor used approximately 28 W, while today the average CRT monitor uses approximately 75 W, or more for larger screens (Roberson et al. 2002). This increase is due to a shift in the 1990s towards the almost universal use of colour screens and larger screens with higher resolution, which use more energy (Koomey et al. 1995). Norford et al. (1988) reported that in 1986 monochrome monitors outnumbered colour monitors by nearly two-to-one.

Standby power consumption for monitors has been steadily decreasing as power management tools become more efficient. The power reduction targets first set by the Energy Star program when it was introduced in 1992 were less than 30 W in standby mode (Norford and Dandridge 1993), but as the technologies have improved, monitors must power down to less than 15 W in standby mode to qualify for Energy Star (EPA 2002). Power-managed monitors now have standby modes drawing as little as 1 W for 14- to 17-inch screens and 2 W for 19- to 21-inch screens (EERE 2000).

It should be noted that monitor power varies according to the image displayed, the monitor's resolution, and other settings. The University of New South Wales found that a typical monitor that may use 60 W when the screen is black requires nearly 80 W to display a full white screen with maximum brightness and contrast (UNSW 2000b). Roberson et al. (2002) identified a similar trend. Wilkins and Hosni (2000) found that monitors displaying Windows used more energy than monitors displaying DOS, most probably due to the reasons noted above.

3.2.1 Flat-Panel Displays (FPDs)

In the 1990s, although computer displays tended towards larger, higher-resolution colour monitors, several emerging flat-panel display (FPD) technologies offered some relief to the historical increase in power (Nordman et al. 1997). Standard coloured liquid crystal displays (LCDs) use less than 25 W in active mode (Figure 4). A preliminary study by Roberson et al. (2002), however, found that power use per square inch of display area increases with the LCD size, unlike CRTs. The average power requirements of three 18-inch LCDs measured was 54 W. Roberson et al. suggested that larger LCDs shouldn't use any more power per unit display area than smaller LCDs, and therefore potential exists to decrease power requirements of larger LCDs in future models. Monochrome LCDs require approximately four times less power than coloured LCDs.

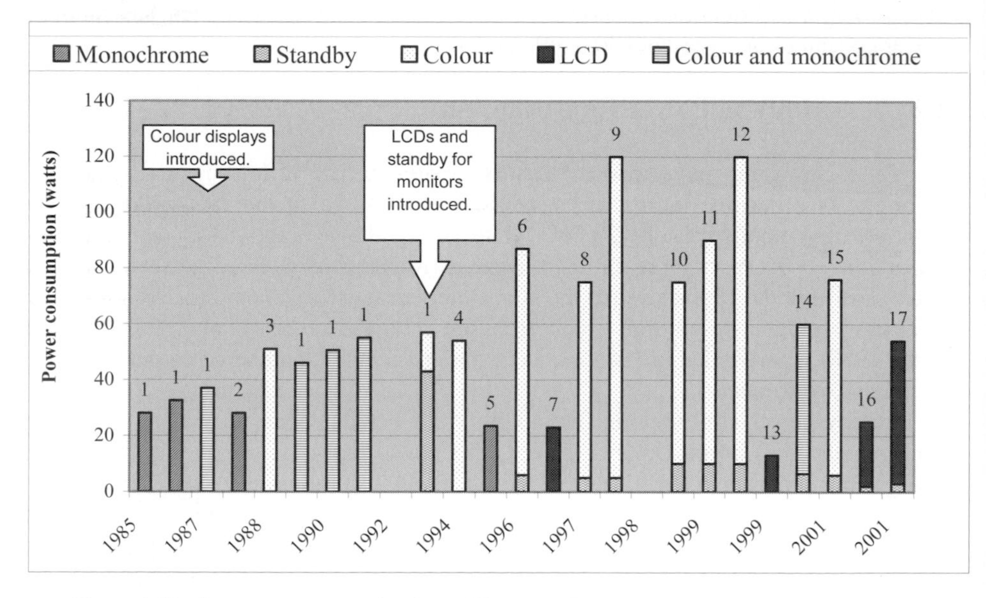

Figure 4: Monitor power consumption from different studies, according to year.
Sources: Numbers above the bars refer to the source of the data. The figure should be read with caution as different types and sizes of monitors are presented.

1. Figures based on measured data by various authors, trade press assessments, personal communications from industry participants (Koomey et al. 1995).
2. Average of 10 monochrome monitors (Norford et al. 1988).
3. Average of 4 colour monitors measured (Norford et al. 1988).
4. Average of 87 colour monitors. Standby mode not recorded (Newsham and Tiller 1994).
5. Average of 25 monochrome monitors (Newsham and Tiller 1994).
6. Average of 3 17-inch colour monitors (Nordman et al. 1996).
7. Colour 10-inch LCD, standby power consumption not specified (Nordman et al. 1997).
8. 15-inch monitor (Nordman et al. 1997).
9. 20-inch monitor (Nordman et al. 1997).

10. Average 15-inch monitor. Figures based on measurements by various authors (Nordman 1999).
11. Average 17-inch monitor. Figures based on measurements by various authors (Nordman 1999).
12. Average 21-inch monitor. Figures based on measurements by various authors (Nordman 1999).
13. Average of 13 15-inch LCDs (OEP 1999). Please note that the average maximum power (according to the manufacturers specifications) of LCDs quoted was calculated to be 39 W. As previously mentioned, the nameplate overstates the actual measured power consumption by a factor of two to four. To allow for unbiased comparison with the other data presented in this figure the average maximum power has been reduced by a factor of three.
14. Based on 613 colour and monochrome monitors (UNSW 2000b).
15. Average monitor sold (Roberson et al. 2002).
16. Average of 13 15- and 17-inch LCDs (Roberson et al. 2002).
17. Average of 3 18-inch LCDs (Roberson et al. 2002).

Nordman et al. (1997) reported that thin CRTs, based on the same tube technology as standard CRTs but using a thin, flat tube instead of the traditional bell-shaped tube, may use as little as 2 W for a standard-sized monitor in active mode. Their growth in popularity, however, remains to be seen, as prices remain exorbitant.

3.3 Laptop Power Consumption

Computing power has increased since the first models of laptops, however, improvements in energy-saving technologies (driven by the desire to increase battery life) have prevented an increase in overall laptop energy consumption. Laptops consume less than 30 W in active mode (Figure 5), although this may be as low as 6 to 7 W in some models (Lorch and Smith 1998). Power consumption decreases to approximately 4 W in standby mode.

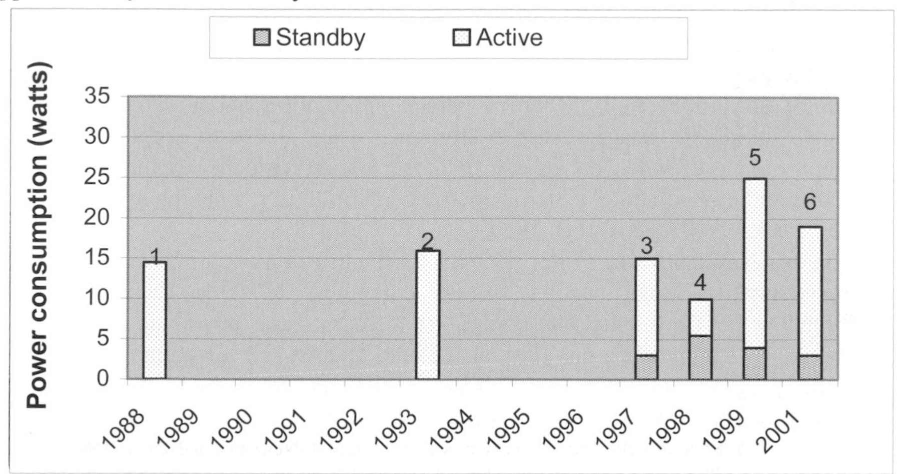

Figure 5: Laptop power consumption from different studies, according to year.
Sources: Numbers above the bars refer to the source of the data.

Notes 1. Average of 4 laptops (Norford et al. 1988).
2. Average of 7 laptops (Norford and Dandridge 1993).
3. Average laptop (Nordman et al. 1997).
4. Average of 3 Apple laptops (Lorch and Smith 1998).
5. Average laptop. Figures based on measurements by various authors (Nordman 1999).
6. Average of 9 laptops (Roberson et al. 2002).

Lorch and Smith (1998) found that for laptops the backlight, processor, and hard drive consumed the most power in active mode. These three components accounted for between 55 and 68 percent of power for three models of Apple laptops. Another study quoted in Lorch and Smith found that 14 percent of total power consumption is due to the backlight, 13 percent to the processor, and 14 percent to the hard drive.

3.4 Minicomputers and Mainframes

The power consumption of minicomputers and mainframes has been significantly decreasing (Figures 6 and 7) as more and more of the functions previously handled by peripherals are integrated into fewer and fewer chips. Koomey et al. (1995) suggest that heat is an especially important issue in these machines (it reduces equipment life), and manufacturers have pushed to reduce energy use for this reason.

3.5 Patterns of Use

While PC power consumption has been considered above, the energy consumption of an individual computer system will also be determined by its patterns of use—how long it is turned on and how long it is actually in use. Prior to the 1990s, little measured data were available to determine patterns of use for computer systems. Norford et al. (1988) said, "There is much speculation about daily profiles for electronic equipment usage...Anecdotally, PC usage patterns range from those consistently shut off when not in use to those left on virtually all the time." Most mainframe and minicomputer systems operated continuously, although some were shut down on weekends (Norford et al. 1988).

In 1994, Newsham and Tiller (1994) used custom-made software to develop desktop computer usage profiles of 94 computers at three Canadian federal government sites. They found that computers were used for only about 12 percent of the time they were switched on. They also found that on average computers were left on overnight 21 percent of the time. Similar results were found by Piette et al. (1995). They specified that a workday is 9.5 hours long, with a typical PC system in "active" mode for four hours and in "idle" mode for 5.5 hours. While only 76 percent of PC systems are on during any given weekday, 18 percent are left on all night for 14.5 hours, and 20 percent are left on all weekend.

PCs are being left on at night with increasing frequency, possibly reflecting the growth of the Internet and network connections (Perry 2001). A study by Webber et al.

(2001) involving more than 1,000 PCs across 11 sites deduced that in the United States 56 percent of computers and 68 percent of monitors are left on at night in offices. Energy audits conducted in 2001 of the 5,000 computer systems in offices at the University of New South Wales in Sydney, Australia found that 43 percent of computers were left on at night. For monitors, 54 percent were left on or in standby mode at night (UNSW 2001).

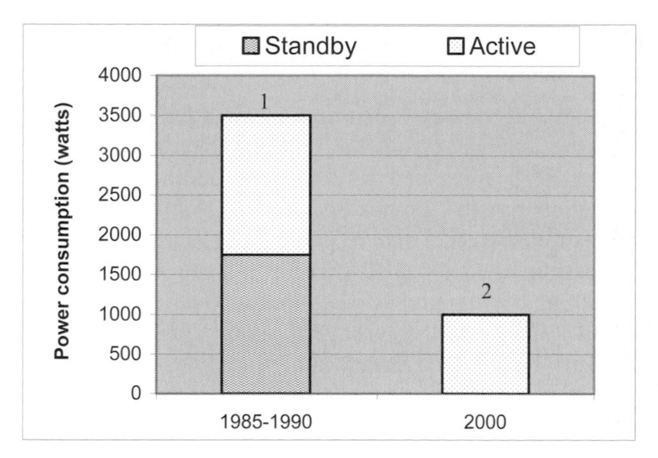

Figure 6: Power consumption of minicomputers from different studies, according to year.
Sources: The numbers above the bars refer to the source of the data.

1. Average model (Koomey et al. 1995).
2. Average model (Kawamoto et al. 2000).

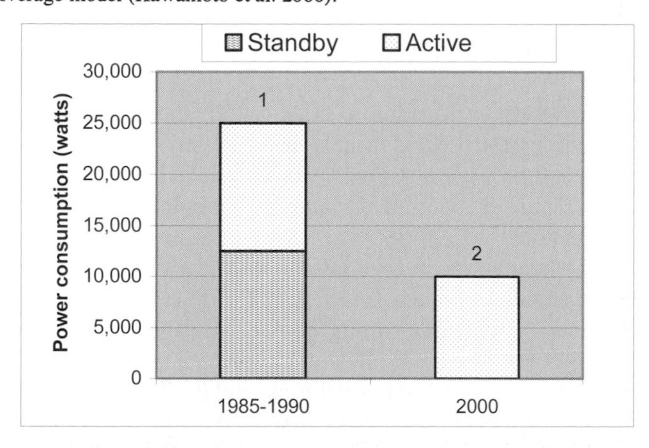

Figure 7: Power consumption of mainframe computers from different studies, according to year.
Sources: Numbers above the bars refer to the source of the data.

1. Average model (Koomey et al. 1995).
2. Average model (Kawamoto et al. 2000).

In industrializing nations, computers may be more likely to be switched off when they are not used. Mungwititkul and Mohanty (1997) reported that in Thailand computers are normally switched off at night and weekends, which is reflected in their findings of 53 percent idle energy loss for a computer system compared with the U.S. standard of 74 percent. These patterns are most likely a reflection of cultural differences, driven by the relative scarcity and cost of resources.

3.6 Effects on Heating, Ventilation, and Cooling

In addition to direct PC energy consumption, when in operation PCs generate heat, which requires further energy to cool the building. Nordman et al. (1997) estimated that for every four kWh consumed by PCs, one additional kWh is consumed by cooling and ventilation systems. Lovins and Heede (1990) estimated that while office equipment accounts for between 5 and 20 percent of electrical loads in office buildings, this load can increase to as much as 50 percent in some PC-intensive buildings when contribution to building cooling load is taken into account.

Yamamoto and Abe (1994) suggest that more often than not computer room temperatures are set lower than necessary at around 23 to 24 Celsius (73-75 °F) when computers are in operation. They reviewed technical and reliability aspects with eight leading computer systems manufacturers. The results led Yamamoto and Abe to conclude that temperatures could be increased to an upper limit of 28 Celsius (82 °F) without negatively affecting the computer equipment, providing the room temperature was not increased in a short time and that condensation could be prevented.

4. PC ENERGY MANAGEMENT

A comprehensive study by Kawamoto et al. (2000) found that power management for office equipment in the United States in 1999 saved 23 TWh per year, and complete saturation and proper functioning would have achieved additional savings of 17 TWh per year (Figure 8). Complete saturation of night shutdown would reduce power use by an additional 7 TWh per year. Many of these potential savings would be achieved by desktop computers and monitors, because the proportion of PCs with power management enabled is much lower than for other types of office equipment, and also because power reductions through power management are large (especially for monitors).

Significant energy savings can be achieved by switching a computer and monitor off and by utilizing automatic energy management features. Nordman et al. (2000) estimated that when left on at night, a computer and monitor with the energy management features disabled use approximately six times the energy used by a computer and monitor with the automatic energy management features activated and that are switched off at night, as illustrated in Figure 9. These estimates are based on a computer and monitor that consume 45 W and 90 W when in active use, 25 W and 5

W when in standby mode, and 2 W each while switched off. The PC has 9.5 hours of on-time during the workday and the user does not turn on their PC for 20 percent of workdays (due to absenteeism, meetings, travel, and vacations). "Disabled/enabled" refers to power management functioning on both the computer and monitor. "On," "low," and "off" refer to the night status of the computer and monitor, respectively ("low" indicates that the equipment is in a low power state). The last bar in the figure reflects the best estimate of typical office night status.

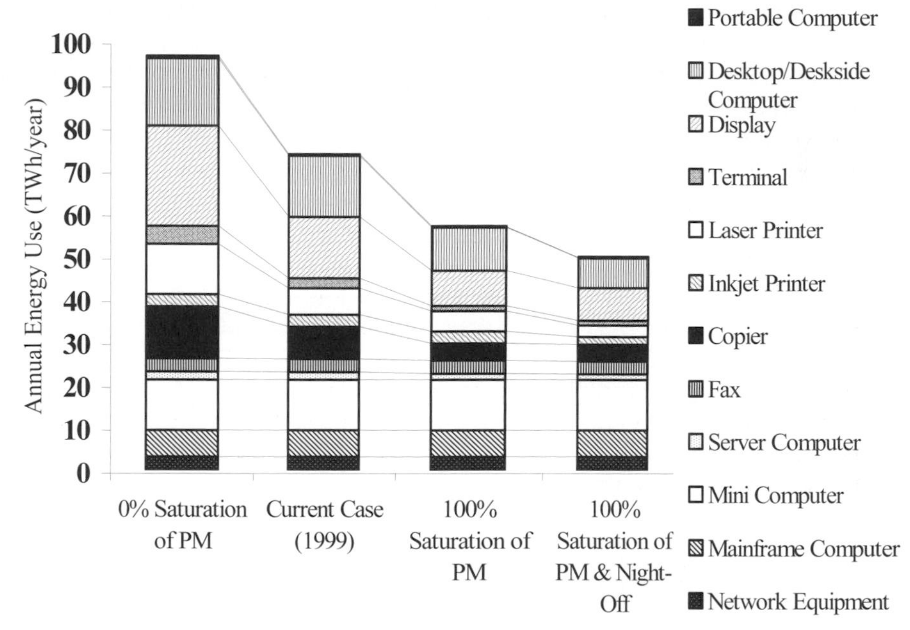

Figure 8: Energy use by equipment type as a function of power management levels.
Source: Kawamoto et al. (2000).

Automatic energy management features are now standard in most PCs, but more often than not these features are either not enabled or not correctly enabled. Automatic energy management features, such as Energy Star, are discussed below, including problems which were prevalent in earlier power-managed PCs. In most cases, however, automatic power management is not a substitute for switching the machine off when not in use for extended periods of time. Substantial energy savings can also be made by switching the PC off, since it has been found that more than 50 percent of computers are left on at night (see section above). Unfortunately, there is a common misconception that switching a PC on and off will harm it. In this section we also examine the evidence, both for and against switching.

4.1 Automatic Power Management

Automatic power management saves energy by putting the computer and monitor into a low power mode during periods of inactivity by temporarily reducing their speed or functionality. Built-in power management was first introduced in laptop computers to extend battery life. In the early 1990s, it was brought into the desktop market. Initially, the majority of power management software was only used for monitor power management, not for controlling the computer itself. This was the case with Microsoft operating system software until the release of Windows 95. Many earlier power management systems for both monitors and computers had long recovery times, awkward configuration methods, and low energy savings, but power management has improved rapidly, becoming more reliable, easier to use, and now delivers considerably more energy savings (Nordman et al. 1997).

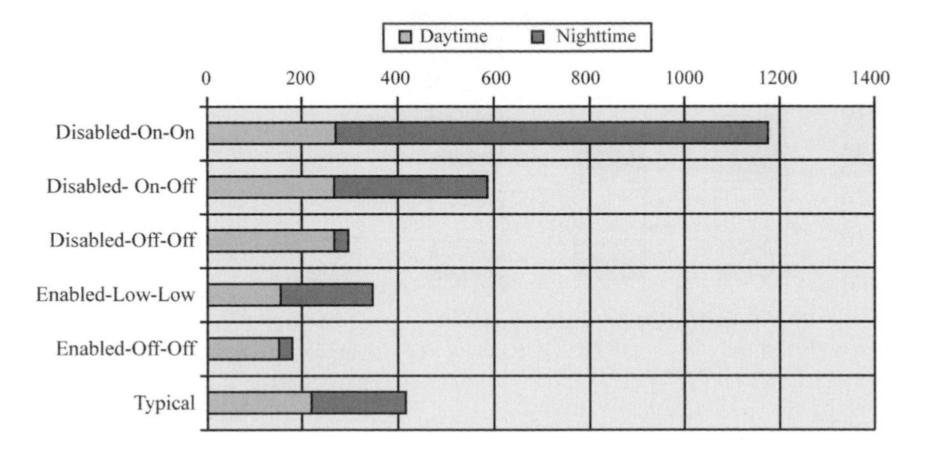

Figure 9: Effects of energy management features and night status on annual energy use of a computer and monitor.
Source: Nordman et al. (2000).

In 1993, Intel and Microsoft first introduced Advanced Power Management (APM), which established an industry standard in power management. APM defines how power management commands are communicated within the PC system. In 1998, Intel, Microsoft, and Toshiba developed the Advanced Configuration and Power Interface (ACPI), where the primary control of power management shifted from the BIOS (basic input/output system) to the operating system (Roberson et al. 2002), thus making power management features more user friendly. ACPI has allowed manufacturers to produce computers that automatically power up as soon as the keyboard is touched. Apple computers use technology that is distinct from but similar to ACPI (Nordman et al. 1996)

Power management interacts with every part of the PC—the operating system, software, CPU, monitor, network cards, video cards, peripherals, etc. Figure 10 demonstrates the communication paths between the components, as described by Nordman et al. (1997). The BIOS sends periodic signals to the operating system to begin power management (2). The operating system checks the application software for activity (3). If no activity is detected, the operating system sends a message back to the BIOS (4), which begins a timer. The BIOS continues to monitor keyboard and mouse activity (1). After a specified time with no activity, the BIOS will initiate power management by sending messages to the hard disk, peripheral cards, processor, and video card (5). After initiating a change in mode, the BIOS begins another timer which indicates when to initiate the next lower power management mode. The BIOS continues to monitor keyboard, mouse, and network activity. If activity is detected, the BIOS will send the appropriate messages to return the PC to an active state.

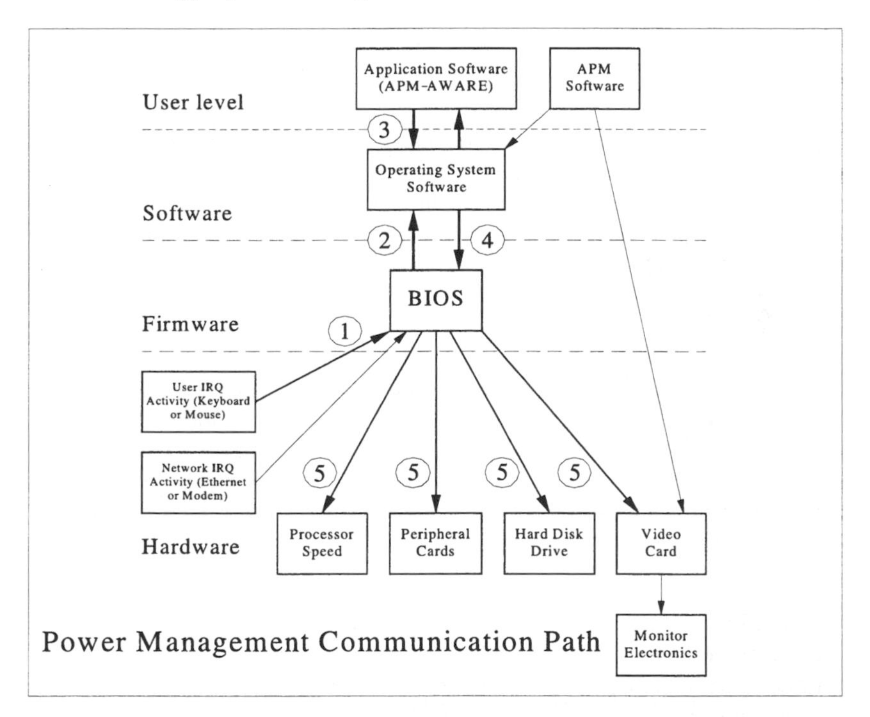

Figure 10: Power management key components and communication paths. *Source*: Nordman et al. (1997).

Many of the components have their own timers, which maximize power management by waking up only the necessary components. In many systems, the hard disk has an independent timer, since it only needs to become active with demand for hard disk activity (Nordman et al. 1997). It is also possible for the monitor to be powered down, even if the computer is not, although the signal to do so must come from the computer. Most monitors are controlled through display power management signaling (DPMS). When the timer indicates that the monitor should be put to sleep, the BIOS (or software) signals the video card, which in turn sends the appropriate DPMS signal to the monitor (Figure 10).

The power management modes for a typical computer system with APM are shown in Table 3. The timing of the power management modes is determined by the settings (usually in the BIOS), specifying the delay between each power management mode and the next. Each successive power management mode represents a decrease in energy consumption and CPU function, and therefore more time is required to bring the computer back to active mode (Nordman et al. 1997). Different PCs will have different power management modes and the same mode may not necessarily represent the same level of activity. Efforts are now underway to standardize power management terminology (Nordman et al. 2002).

4.1.1 Office Energy-Efficiency Schemes

Several industrialized countries have set up policies and implemented schemes and programs to encourage energy efficiency in office equipment. Such schemes include the U.S. EPA's Energy Star program, the Swiss E2000 program, and the Swedish Nutek program. Thailand is the first developing nation to consider energy-savings potential for office equipment, and has introduced an Energy Conservation Promotion Act (ECPA) that makes energy conservation mandatory in high energy-consuming buildings (Mungwititkul and Mohanty 1997).

Energy Star is one of the most successful programs and is emerging as an international standard. Representatives from the U.S. EPA, the European Commission, and the Japanese Ministry of International Trade and Industry (MITI) began negotiations in 1993 to develop a single logo and uniform criteria. In October 1995, the EPA and MITI agreed to co-operate. No formal agreement has been reached between the EPA and the European Commission, although negotiations continue (Howarth et al. 2000). The governments of Australia and New Zealand have adopted the Energy Star label.

4.1.1.1 Energy Star

In 1992, the EPA introduced the Energy Star voluntary labeling program in the United States, which was designed to identify and promote energy-efficient products in order to reduce carbon dioxide emissions. In 1996, the program became operated jointly by the EPA and the Department of Energy. Initially, energy-saving features in monitors, computers, and printers were promoted. The Energy Star label now covers more than 30 consumer product categories, including refrigerators, air conditioners, and dishwashers (Webber et al. 2002).

Table 3: PC power management modes.

Mode	Description	Energy savings		Recovery time (seconds)
		Computer	Monitor	
Full-on	All components fully powered; no power management occurring.	0	0	0
APM enable	CPU is slowed or stopped (depending on BIOS); all other devices still draw full power. Some systems have a "doze" mode that is similar to APM enabled.	0–25%	0	0
APM standby	CPU may be stopped, depending on operation or activity; most devices are in low power mode. Monitor enters its first power management mode. Activity can trigger a return to enabled or full-on, depending on the system and activity.	20–30%	60–90%	n/a
APM suspend	CPU is stopped; most power-managed devices are not powered (network card may stay on). Maximum savings under APM. Activity can trigger a return to standby, doze or full-on, depending on the BIOS.	25–45%	0–10%	3-10
Hard disk power down[1]	Hard disk spin is stopped; this is independent of other power management (hence not a system mode), so that the reminder of the system can be fully operational or powered down. Disk control electronics are still powered to facilitate quick reactivation.	10% (disk savings, independent of other savings)	-	3–10
Hibernate[1]	All memory contents and system state saved to disk. System resistant to power loss.	90–100%	-	15–60
Off	No operational parameters are saved. System resets and starts at full-on mode.	96%[2]	98%[2]	Not specified

Source: Nordman et al. (1997).

Note 1: This is not an APM.

Note 2: Figures obtained from UNSW (2000b) as not specified in Nordman et al. (1997). A switched-off PC that remains plugged into the main electrical supply may still consume energy. UNSW (2000b) found that some computers pulled up to 5 W when switched off but plugged into the socket with the switch on. Roberson et al. (2002b) found similar trends. For maximum savings to be realised, equipment should be switched off at the socket.

The Energy Star program has met with considerable success since its inception, especially in the office equipment market. This was largely due to President Clinton issuing Executive Order 12845, requiring that the U.S. government (the world's largest purchaser of equipment) purchase Energy Star computers, monitors, and printers (Brown et al. 2000). Webber et al. (2002) estimate that 95 percent of monitors and 90 percent of computers sold in the United States are Energy Star-compliant, but in order for energy savings to be realised, the Energy Star functions must be activated.

Nordman (1999) estimates that 80 percent of monitors and 50 percent of computers are Energy Star-activated. The Energy Star program has realised this limitation, and subsequently, manufacturers are now required to ship equipment with the Energy Star features enabled (Webber and Brown 1998).

Energy Star-compliant computers must enter a low power "sleep" mode of less than 15 percent of maximum power usage within 30 minutes of inactivity. For example, a computer that uses a 200-watt power supply must power down to 30 watts or less in low power mode. Energy Star-compliant monitors must power down to a "sleep" mode of 15 watts or less after a 15 to 30-minute period of inactivity. If idle for more than 70 minutes, monitors must enter a "deep sleep" mode of 8 watts or less (EPA 2002). Energy Star is not available for mainframes and minicomputers, despite discussions with several manufacturers in 1997 to develop a version of the Energy Star program for larger computers (Nordman et al. 1997).

4.1.2 Problems with Power Management

Power management has been more successful in monitors than computers. "Compared to power managing [computers], monitors are usually simpler, have much more energy-savings potential, power-manage more reliably, and are less likely to interfere with operation or network connections" (Nordman et al. 1997).

The difficulties in properly configuring power management in computers and monitors are the largest barrier to achieving energy savings from automatic energy management. While Nordman (1999) estimates that 50 percent of computers are Energy Star-activated, as mentioned above, Kawamoto et al. (2000) says that only 25 percent are correctly power-managed to achieve maximum energy savings. A study by Nordman et al. (1996) of power management features and configuration of Energy Star-compliant machines found only 11 percent of CPUs fully enabled and about two-thirds of monitors successfully power-managed. The presence, enabling, and success of power management are matters of degree, rather than the simple presence or absence on individual devices.

It is very difficult to determine whether power management is properly operating in machines. The only indication to the user that power management is occurring in their computer is when the hard disk audibly spins down or delays in the appearance of keystrokes when spinning up. Other than this, it is difficult to know whether the computer is accomplishing any further power management (Nordman 1996). Power management in monitors is more obvious, as the screen will dim or go blank. However, there is still potential for a blank screen-saver to be confused with a power-managed monitor.

Users often believe their equipment is power-managed because the Energy Star logo appears during startup. Many users do not realise that they must first activate the energy management features for energy savings to occur[1]. This is becoming less of a

---

[1] Energy Star features can be activated through options on the desktop.

problem now that it is compulsory for manufacturers to ship Energy Star-compliant products with the feature activated (Webber and Brown 1998).

Users may also find that although the Energy Star logo is present, their computer system may not be capable of energy management, due to its operating system. For example, Microsoft's Windows NT (Version 4 and earlier) is incompatible with Energy Star and will prevent power management (for both computers and monitors) from operating.

Some computer systems that are not Energy Star-compliant may still be capable of energy management. Often, there is no obvious indication of power management features, and so the potential energy savings may never be realised (Koomey et al. 1995).

Once enabled, power management may present some further challenges. Power management interacts with every part of the computer, and therefore there is potential for unexpected interactions, which may cause problems. While these problems occur infrequently and usually only with older equipment, when they do occur they tend to generate considerable work for support staff. The memory of such problems may cause support staff to routinely disable power management (Picklum et al. 1999).

4.1.2.1 Network Complications

Power management acting on some combinations of hardware and software has been known to cause loss of network connections, although this was more of a problem for earlier systems (Nordman et al. 1996, 1997). In many networks, a central server will send out periodic "Are you there?" messages to determine which computers are still on and connected to the network. These messages may cause enough activity to keep the computer and monitor awake, thus energy savings are not realised. If the PC does go to sleep, it may fail to respond to the "Are you there?" message, and therefore the network connection may be dropped (Nordman et al. 1997).

A growing number of BIOS are able to differentiate network activity from other activity, such as keyboard activity, so that only those parts of the system needed to respond to the network will be powered up. Instantly Available PC (IAPC), such as Microsoft's OnNow initiative, represents one possible direction for the future of power management, where the PC stays on continuously so that the PC is ready to use immediately when the user presses the On button. The PC is perceived to be off when not in use, but is still capable of responding to wake-up events from networks, etc. Microsoft is particularly interested in developing these features for home computers, which they believe require instant accessibility. They are also interested in introducing OnNow features into all PCs, whether they are laptops, desktops, or servers (Nordman et al. 1997).

4.1.2.2 Software Interactions

Some application software can interfere with power management. For example, the "auto-save" function in word processors and spreadsheets may cause the PC to wake up if auto-save occurs even when there have been no changes, defeating power management entirely or partially. Some screen-savers may periodically load complex images from the hard disk, preventing the disk from powering down (Nordman et al. 1997).

4.1.2.3 Re-activation Times

There are problems and inconveniences associated with the time taken for computer equipment to return to an active state from a low power state. Computers activated with the first generation of power management software sometimes took more than 30 seconds to return to the active state. With IAPC, manufacturers can now design computers that go into extremely low power modes (less than 5 watts) while awakening instantaneously (EPA 2002).

4.1.3 After-Market Devices

For computers and monitors which do not have built-in power management features, devices are available that sense activity and reduce energy usage where appropriate. The add-on device is connected externally to the computer or monitor and senses keyboard/mouse activity or the presence of a person in front of the computer. Some add-on devices work with software that will save any open documents and return to the file previously being worked on when the machine is powered back up (Nordman et al. 1997).

Adaptor cords also exist that ensure that monitors without built-in power management capabilities are off whenever the computer has been switched off (Nordman et al. 1996).

4.2 Switching PCs Off

Maximum energy savings result from manually switching the PC off when not in use, rather than relying on power management features alone (Figure 9). Switching PCs off when not in use has further benefits over leaving them running continuously. Many computers perform diagnostic tests on start-up, and never power-cycling (i.e., turning on and off) the computer means that these tests are not performed on a regular basis (Koomey et al. 1993). Leaving PCs on at night may also pose a fire hazard (Gateway 2001). Hackers or lurkers in the night cannot access a computer that is switched off. Less energy will be required to air-condition the office, due to less heat being produced by equipment overnight. And, of course, a PC that is switched off saves money on electricity bills.

Koomey et al. (1993) found that shutting computers off for several hours is not a cost-effective use of peoples' time; however, overnight and weekend shutdown is economically advantageous. A computer which typically requires 15 seconds of the user's attention per day to startup and shutdown, 5 days per year, 50 weeks per year, causes the user to spend about one hour per year in tasks that are essentially unproductive. (Startup and shutdown will take considerably longer than 15 seconds per day; however, many users perform other tasks during this time.) The greatest savings in terms of energy and labour costs can be achieved by relying on power management features during the day when the computer is not in use and shutting down and switching off the computer at night. Switching off the monitor during the day, however, is sensible, as this action takes little time and can save a substantial fraction of the PC's energy use (Koomey et al. 1993).

Many organizations have adopted PC switch-off policies and implemented campaigns with impressive results. However, any switch-off campaign must be comprehensive and ongoing to significantly reduce energy consumption. Newsham and Tiller (1994) found that stickers placed on monitors reminding the user to switch off PCs when not being used produced reductions in mean PC energy consumption of 14 percent over an eight-week period, but these savings diminished with time. By comparison, the activation of an automatic power management system produced reductions in mean computer energy consumption of 63 percent, while the monitor consumption was reduced by 82 percent. These savings were maintained over an eight-week period.

Many users are reluctant to switch off computers at night or when not in use, due to the common view that regular power cycling will harm equipment. Energy managers often claim that power cycling does not damage equipment; however, these claims are seldom supported by objective evidence. Research suggests that switching computers on and off or continuous operation both have drawbacks (more so for computers greater than five years old) in terms of absolute computer life span, but the probability of failure over the "operational" life is negligible, as discussed below.

4.2.1 Thermal Cycling

Switching a computer on and off causes thermal cycling, which may give rise to mechanical failure. Mechanical failure from thermal cycling occurs when materials with different expansion co-efficients are bonded together. This includes integrated circuit boards (IC), wire-bond assemblies, and printed circuit board (PCB) solder joints. ICs in most desktop computers are plastic-encapsulated micro-circuit 1 (PEM) types. PEMs can develop stress cracks and solder joints can fail due to the metal hardening from the repeated stress-causing cracks in the joint. A theoretical study considered mini-thermal cycles due to partial power cycling. It was found that predicted lifetime was halved from eleven years with no mini-cycles to five years with 20 mini-cycles per day (Mercer et al. 1997).

4.2.2 Hard Disk Drive Effects

In older computers, where the hard disk was frequently power cycled, the disk was prone to more wear due to an increased distance of sliding contact. This occurred when the normally free-flying heads landed and took off as the disk spindle stopped and started (Mercer et al. 1997). But Nordman et al. (1997) and Woolfe (2000) say that hard disks no longer automatically park their heads when shut off, and therefore frequent power cycling is not a problem for the modern hard disk. The University of Tasmania (UTAS 2001) says that modern hard drives are designed and tested to operate reliably for many thousands of hours, including thousands of on/off cycles.

4.2.3 Power Supply Start-up Transients

When the power supply is first turned on, there is a delay while the power supply control circuits establish their correct working conditions. During this period, it is possible for the output voltage to exceed the normal value before full regulation is achieved, giving rise to a "turn-on voltage overshoot." Many modern switch-mode power supplies have soft start control circuits that prevent overshoot, avoiding damage to the components (Mercer et al 1997).

4.2.4 Mechanical Wear

The modern hard disk is less likely to fail if it runs for shorter, rather than longer periods. Hard disks and cooling fans contain moving parts, and are therefore prone to wear with time. Modern hard disks are specified with a mean time to failure of at least 500,000 hours for the first five years of operation, which corresponds to 1.7 failures per year. Beyond the first five years, the failure rate rapidly increases. The lifetime of a typical fan is specified as 50,000 hours (5.7 years) (Mercer et al. 1997). A compounding consequence of fan failure is overheating, which reduces the life of other components in the system (Mercer et al. 1997; Linc IT 1995).

Desktop computers usually don't filter their cooling air. Dust build-up on internal components is related to operating time. Leaving computers on for extended periods of time can increase internal operating temperatures and accelerate failure (EPA 2002; Mercer et al. 1997).

4.2.5 Power Surges

Extending the time a machine is left on increases the probability of damage due to power surges and spikes caused by the loss of large loads and everyday switching of inductive loads in the power system (EPA 2002; Mercer et al. 1997).

It is a misconception that switching on multiple PCs in the morning will cause a power surge, unless they are switched on at precisely the same time, which is very unlikely unless they are connected to the same switch. Air-conditioning and lighting use much more electricity than PCs in offices on a square-foot basis, and therefore the turning on of these electrical appliances is a far greater spike in demand than PCs (Koomey 2001). Soft switches may enable networked computers to be turned off and on remotely. Presumably, if all computers in a large office were switched on at the same time via a central switch, a power surge could occur, especially in a poorly-designed circuit. Systems managers need to ensure that the start-up of PCs is staggered, should they wish to reduce PC energy consumption by establishing a central switch on/off mechanism.

4.2.6 Switching Monitors Off

Unlike computers, there is very little debate as to whether or not to switch monitors off when not in use. Monitors are similar to TV sets and no one would suggest that they should be left on when not in use. The University of New South Wales (UNSW 2000a) contacted seven major monitor manufacturers, all of whom agreed that turning the monitor off when not in use is the best way to extend its life and save energy. The U.S. EPA (2002) reported that switching the monitor on and off five times or more a day increases the frequency of faults in power transistors in the control and deflection parts only after the machine has been used 20 to 30 years. A typical monitor would become technologically redundant before it fatigued due to frequent switching.

As can be seen from the information, maximum cost and energy savings can be achieved by switching off monitors whenever they are not in use, even if the computer must be left on. Energy features should also be activated to ensure that the monitor enters a low power state if the user forgets to turn it off. As for computers, both power cycling and continuous operation have the potential to reduce the lifespan, although the modern-day computer is designed to better deal with frequent power cycling. It is recommended that computers are switched off at night and weekends and that power management features are utilized during the day. The majority of computer manufacturers now recommend that computers be switched off at the end of the day and on weekends (Mercer et al. 1997; Waghorn 2001; EPA 2002; Gateway 2001; and UNSW 2000a). This action will not appreciably affect computer longevity and may also generate substantial energy and cost savings

5. CONCLUSION

Office equipment consumes three percent of the U.S.'s energy requirement, more than 40 percent of which can be can be attributed to the operation of PCs. Growth in energy requirements for office equipment is beginning to slow as the market becomes saturated (especially for computers and monitors) and as equipment becomes more energy-efficient. More and more companies are also developing and implementing strategies to reduce their greenhouse gas emissions. Major energy service companies are increasingly offering "energy outsourcing" deals in which they take over corporate energy management and invest in energy-efficiency. These deals eliminate many of the barriers that have slowed more widespread adoption of energy-efficiency technologies and strategies in the commercial sector in the past decade (Romm et al. 1999).

While standard monitors and, more recently, computers, require more power in active mode than previous models, there have been significant improvements in standby power requirements. The further adoption of energy-efficient hardware technologies is also likely to decrease power requirements of the most commonly used PCs in the future. FPD technologies that are extremely energy-efficient are beginning to replace CRTs as the technology improves and prices fall. Koomey (2001) predicts that the changeover to FPDs will probably be complete in the market by 2010, though it may take five additional years for all existing CRTs to be displaced after 2010. Laptop computers, which consume considerably less power than a desktop computer and monitor, are also increasing in popularity. Thin clients look promising for future PC energy-efficiency in the commercial sector.

Computer manufacturers have addressed many of the problems associated with PC power management by making the feature more flexible and more compatible with current PC networks. Power management now has many more options, which allows specific problems to be addressed without disabling all power management features. As the technology has matured, power management has become an effective tool for saving energy; however, a significant proportion of PCs are not correctly power management-enabled, and therefore maximum savings from power management technology are yet to be realised.

Despite improvements in power management features and the gaining popularity of more energy-efficient hardware technologies overall, PC energy requirements are expected to continue to increase. A significant contributing factor is that an increasing proportion of PCs are left on overnight unnecessarily. Many users are reluctant to switch off PCs when not in use due to the common view that regular power cycling will harm equipment. Both power cycling and continuous operation have the potential to reduce equipment life, although the modern computer is designed to better deal with frequent cycling. The majority of manufacturers now recommend that computers be switched off at night and on weekends and that monitors be switched off when not in use.

Misconceptions and a lack of awareness on the side of the user are obviously key obstacles in realizing maximum PC energy savings, which suggests that education of users and systems managers is critical. Firms and government agencies need to respond by ensuring that the implementation of policy clearly identifies and puts to rest myths and misconceptions which prevent the user from taking steps to reduce PC energy consumption.

PCs are one of the fastest changing technologies in our society. Not only are the hardware and software changing, but the Internet and other emerging information technologies continue to introduce new and expanded uses for the not-so-humble PC. These technological advances directly impact the trends in PC energy consumption. While this chapter places greater emphasis on PCs in the commercial sector, the impact of PCs in the residential sector must not be overlooked. The proliferation of PCs in the home, due to expanded use of the Internet, means that the residential sector may be responsible for a much greater proportion of energy consumed by office equipment than previous estimates. Re-estimates of PC energy consumption, both in the commercial and residential sectors, will be required in the near future due to new technological developments and the changing ways in which we use PCs.

REFERENCES

Brown, R., C. Webber and J. Koomey. 2000. Status and future direction of the Energy Star program. Proceedings of the ACEEE Summer Study, August, at Asilomar, CA. U.S. Department of Energy, U.S. EPA. <http://enduse.lbl.gov/Info/Pubs.html> (14 January 2003).

Energy Efficiency and Renewable Energy (EERE). 2000. How to buy an energy-efficient computer monitor. Office of Energy Efficiency and Renewable Energy, U.S. Department of Energy. <http://www.eren.doe.gov/femp/procurement/pdfs/monitor.pdf> (14 January 2003).

Environmental Protection Agency (EPA). 2002. Energy Star Web site. http://208.254.22.7/index.cfm?c= home.index> (23 May 2003).

Gateway. 2001. Should a computer run continuously? <www.gateway.com/support/faq/c-gwpc/g040400. shtml> (14 January 2003).

Greenbergy, S., C. Anderson and J. Mitchell-Jackson. 2001. *Power to the people: Comparing power usage for PCs and thin clients in an office network environment.* Scottsdale, AZ: Thin Client Computing.

Howarth, R., B. Haddad and B. Paton. 2000. The economics of energy efficiency: Insights from voluntary participation programs. *Energy Policy* 28:477–486.

Huber, P. 1999. Dig more coal—the PCs are coming. *Forbes Magazine*, 5 May.

Japan Electronics Industry Development Association (JEIDA). 1999. *The white paper of personal computer 1999–2000.* Tokyo: JEIDA.

Kawamoto, K., J. Koomey, B. Nordman and A. Meier. 2000. Electricity used by office equipment and network equipment in the U.S. Conference: Energy Efficiency in Buildings. Berkeley, CA: U.S. EPA, Lawrence Berkeley National Laboratory. <http://enduse.lbl.gov/Info/Pubs.html> (14 January 2003).

Koomey, J. 2000. Rebuttal of Mills' congressional testimony. Berkeley, CA: Lawrence Berkeley National Laboratory. <http://enduse.lbl.gov/projects/infotech.html> (14 January 2003).

———. 2001. Personal communication. Berkeley: Lawrence Berkeley National Laboratory.

Koomey, J., T. Oey and E. Bergman. 1993. The economics of cycling personal computers. *Energy Policy* (September): 937–943.

Koomey, Y., M. Cramer, M. Piette and J. Eto. 1995. Efficiency improvements in U.S. office equipment: Expected policy impacts and uncertainties. Berkeley, CA: Lawrence Berkeley National Laboratory, U.S. Department of Energy.

Laitner, J., J. Koomey and E. Gumerman. 2000. Re-estimating the annual energy outlook 2000 forecasting using updated assumptions about the Internet economy. Berkeley: U.S. Department of Energy, U.S. EPA, Lawrence Berkeley National Laboratory. <http://enduse.lbl.gov/projects/infotech.html> (14 January 2003).

Linc IT. 1995. *PC shutdown—To be or not to be.* Sydney: Commonwealth Department of Primary Industries and Energy.

Lorch, J. and A. Smith. 1998. Apple Macintosh's energy consumption. *IEE Micro* 18(6) (November–December): 54–63. Lawrence Berkeley National Laboratory.

Lovins, A. and H. Heede. 1990. *Electricity saving office equipment.* Snowmass, CO: Rocky Mountain Institute.

Matthews, S. 2001. Personal communication. Pittsburgh, PA: Department of Engineering and Public Policy, Carnegie Mellon University.

Mercer, K., V. Kasula and R. Nilson. 1997. Does power cycling a desktop computer harm it? A review of the literature. New Zealand Energy Efficiency and Conservation Authority. <www.energywise.co.nz/content/ew_business/Office_Equipment/Power_cycling.html> (June 2001).

Mills, M. 1999. *The Internet begins with coal.* Greening Earth Society, Mills-McCarthy and Associates, Inc.

Mungwititkul, W. and B. Mohanty. 1997. Energy efficiency of office equipment in commercial buildings: The case of Thailand. *Energy* 22(7): 673–680.

Newsham, G. and T. Tiller. 1994. The energy consumption of desktop computers: Measurement and savings potential. *IEEE Transactions on Industry Applications* 30(4) (July/August): 1065–1070.

Nordman, B. 1999. Power levels, operating patterns, and energy use. Berkeley, CA: Lawrence Berkeley National Laboratory. <http://eetd.lbl.gov/bea/sf/> (14 January 2003).

Nordman, B., M. Piette and K. Kinney. 1996. Measured energy savings and performance of power-managed personal computers and monitors. Proceedings from Energy Efficient Economy Summer Study.

Nordman, B., M. Piette, K. Kinney and C. Webber. 1997. User guide to power management for PCs and monitors. Berkeley, CA: U.S. EPA, Lawrence Berkeley National Laboratory. <http://eetd.lbl.gov/EA/Reports/39466/> (14 January 2003).

Nordman, B., A. Meier and M. Piette. 2000. PC night status: Power management enabling and manual turn-off. Berkeley, CA: Lawrence Berkeley National Laboratory.

Nordman, B., A. Meier and D. Aumann. 2002. Even electronics deserve a good night's sleep: Towards standard power user interface elements. Proceedings of the 2002 ACEEE Summer Study on Energy Efficiency in Buildings, 18-23 August, at Pacific Grove, CA (in press). Washington, D.C.: American Council for an Energy Efficient Economy.

Norford, L., A. Rabl, J. Harris and Y. Roturier. 1988. Electronic office equipment: The impact of market trends and technology on end use demand for electricity. Berkeley, CA: Lawrence Berkeley National Laboratory.

Norford, L. and C. Dandridge. 1993. Near term technology review of electronic office equipment. Conference record of the 1993 IEEE Industry Applications Society Annual Meeting, at Toronto, 2:1355–1362.

OEP. 1999. Rising performance, falling prices help LCDs push CRTs off desktops. *Office Equipment & Products* 28(248) (May): 17–18, 21. Japan: Dempa Publications.

Peppercorn, S. 2001 Power supplies. *Atomic,* June: 26–28.

Perry, T. 2001. Fueling the Internet. *IEEE Spectrum* 38, Issue 1 (January): 80–83.

Picklum, R., B. Nordman and B. Kresch. 1999. Guide to reducing energy use in office equipment. Berkeley, CA: U.S. Department of Energy, Lawrence Berkeley National Laboratory. <http://eetd.lbl.gov/bea/sf> (14 January 2003).

Piette, M., M. Cramer, J. Eto and J. Koomey. 1995. Office technology energy use and savings potential in New York. LBNL Report Number LBL-36752. Berkeley, CA: Lawrence Berkeley National Laboratory.

Roberson, J., G. Homan, A. Mahajan, B. Nordman, C. Webber, R. Brown, M. McWhinney and J. Koomey. 2002. Energy use and power levels in new monitors and personal computers. Berkeley, CA: Lawrence Berkeley National Laboratory. <http://www-library.lbl.gov/docs/LBNL/485/81/PDF/LBNL-48581.pdf> (14 January 2003).

Romm, J. 2002. The Internet and the new energy economy. *Resources, Conservation and Recycling* 36:197–210.

Romm, J., A. Rosenfeld and S. Herman. 1999. *The Internet economy and global warming: A scenario of the impact of e-commerce on energy and the environment.* Arlington, VA: The Centre for Energy and Climate Solutions, The Global Environment and Technology Foundation. <http://www.cool-companies.org/energy/paper1.cfm> (23 May 2003).

Roth, K., F. Goldstein and J. Kleinman. 2002. *Energy consumption by office and telecommunications equipment in commercial buildings, volume I: Energy consumption baseline.* Prepared by Arthur D. Little. Washington, D.C.: U.S. Department of Energy. <http://www.eren.doe.gov/buildings/documents/> (14 January 2003).

University of New South Wales (UNSW). 2000a. Personal computers and energy conservation—the case for switching off, unswitch. Unpublished report. Sydney: University of New South Wales.

———. 2000b. UNSW library building office equipment energy audit, unswitch. Unpublished report. Sydney: University of New South Wales.

———. 2001. *Unswitch.* Sydney: University of New South Wales. <www.unswitch.facilities.unsw.edu.au> (14 January 2003).

University of Tasmania (UTAS). 2001. Green computing. Information Technology Services, University of Tasmania. <http://www.its.utas.edu.au/documentation/misc/green.computing.html> (14 January 2003).

Waghorn, C. 2001. More people switch-off computers. Energy Efficiency and Conservation Authority, New Zealand. <http://www.energywise.co.nz> (June 2001).

Webber, C. and R. Brown. 1998. Savings potential of Energy Star volunteer labeling programs. Proceedings of the ACEEE Summer Study, Lawrence Berkeley National Laboratory, California. <http://enduse.lbl.gov/Info/Pubs.html> (14 January 2003).

Webber, C., R. Brown, A. Mahajan and J. Koomey. 2002. Savings estimates for the Energy Star voluntary labeling program 2001, status report. Berkeley, CA: Lawrence Berkeley National Laboratory. <http://enduse.lbl.gov/INFO/LBNL-48496.pdf> (14 January 2003).

Webber, C., J. Roberson, R. Brown, C. Payne, B. Nordman and J. Koomey. 2001. Field surveys of office equipment operating patterns. Berkeley, CA: Lawrence Berkeley National Laboratory. <http://enduse.lbl.gov/Info/LBNL-46930.pdf> (14 January 2003).

Webopedia. 2002. <http://www.webopedia.com> (14 January 2003).

Wilkins, C. and M. H. Hosni. 2000. Heat gain from office equipment. *Ashrae Journal* (June): 33–39.

Woolfe, K. 2000. *Computer energy savings.* Berkeley, CA: Lawrence Berkeley National Laboratory Green Team. <http://www.lbl.gov/ehs/wastemin/green_team/compenergy.html> (15 January 2003).

Yamamoto, M. and T. Abe. 1994. The new energy saving way achieved by changing computer culture. *IEEE Transactions on Power Systems* 9(3) (August): 1563–1568.

Chapter 8

PCS AND CONSUMERS—A LOOK AT GREEN DEMAND, USE, AND DISPOSAL

Mohamed Saied and German T. Velasquez
United Nations University, Global Environment Information Centre, Japan

1. CONSUMER POWER AND ITS EFFECTS ON INDUSTRY

Consumer demand plays an important role in environmental sustainability. One aspect of this is the role consumers can play in the potential greening of industry. Through their choices of what to purchase, they can directly influence industry decisions on whether to produce environmentally friendly goods or not. But consumer choice has yet to realize its full potential as a force driving greener production. A variety of reasons can explain this, including cost considerations, the lack of available alternative technologies, and simply unenlightened force of habit. Cost considerations come into play when the environmentally friendly product comes at a higher price—a price that consumers are unwilling to pay. In many cases, there is no available technology that can produce the good for the same price, while avoiding the environmental impacts of concern. The last reason, force of habit, plays a role, even when the green option is economically competitive. The final consumer decision on whether to demand change from industry rests in the balance between their perception of risks and the priorities that they set.

Cost considerations are easy for consumers to understand. Ecological considerations, however, are not as easy, partly because they are inherently more complex than a single number, like a difference in cost, and also because environmental information is often incomplete, misleading—or outright unavailable. If the outcome of balancing these decisions ends simply with cost minimization, green consumerism cannot realize its potential. Many people, however, are willing to make at least a small sacrifice in cost to help preserve the health of our environment. One challenge is to provide consumers with adequate knowledge so that they can make

Computers and the Environment: Understanding and Managing Their Impacts
Edited by Ruediger Kuehr and Eric Williams, pages 161–181.

better decisions when it comes to balancing their priorities between the costs of maintaining a healthy environment and watching out for their pocketbook.

Initially, personal computers (PCs) were viewed as "clean" machinery (Young 1994). Compared to older technologies, microchips and other computer components are created in very "sanitary" environments, which are easily considered as equivalent to being environmentally friendly. This view has changed recently, though, as more evidence accumulates on the environmental impacts of computers and the information age. It is thus relevant to ask whether consumer choice can play a role in making computers greener. Even as computers become an increasingly integral part of everyday activities, consumers still look at them in the same way as they do many other products—they are judged based on the features beneficial to the user. The purpose of this chapter is to examine how environmental information about computers is being communicated to consumers and explore how the effects of green consumer choice can be increased in the future.

The above discussion relates to how consumers affect environmental impacts through their choice of which computer model to *purchase*. The behavior of the consumer is also key in how the computer is *used* and *disposed of*. Green use and disposal of computers could be even more environmentally important than the choice of model. Chapter 3 and 9 argue that extension of lifespan of computers is important for reducing environmental burdens, while Chapter 7 touches on how consumer behavior affects electricity use. Consumer choices of how long to keep their computer, use power management functions, and what to do with the machine at its end-of-life are thus very important for the environment.

2. ECO-LABELING

Eco-labels are a primary tool available to inform consumers about the environmental characteristics of products. They appear as a label or logo that gives consumers actual data on the product and/or lets them know that it meets a fixed set of environmental criteria. The premise is that consumers will, hopefully, prefer products with eco-labels over ones without.

Eco-labels are often characterized in three varieties: types I, II and III. A type I eco-label is essentially a "certificate of approval" for the product given by a third-party organization, such as a governmental agency. These organizations set a variety of standards that a product must meet in order to use the label, and companies apply to them for the right to put it on their products. In the case of a type II eco-label, a company declares that its product meets independent standards—such as for recyclability or energy-efficiency. A type III eco-label is designed to provide a set of quantitative environmental data to consumers (such as electricity and/or water use) so that the consumers can use this information themselves to evaluate products. Guidelines on eco-labels are included as part of the ISO 14000 series of environmental management standards.

In considering eco-labels it is also important to distinguish between product-dependent and process-dependent information. Product-dependent information deals with the performance and composition of the product itself, such as energy consumption and the presence of undesirable substances (heavy metals, for example). Process-dependent information relates to the process used to create the product, such as use of chemical fertilizers or pesticides in growing vegetables or use of climate change-inducing perflourocarbons (PFCs) in semiconductor fabrication. The certification of produce as "organic" is a process-based eco-label most consumers are aware of. When incorporating process-dependent aspects into an eco-label, the credibility of the label will be enhanced if a third-party organization can verify that processes are being carried out as advertised.

2.1 Environmental Issues for Computers and Eco-labeling

Eco-labeling of PCs poses a number of difficulties. One issue is that PCs are complex and rapidly changing products—a typical PC consists of 1,000 to 1,500 components and 2,000 to 3000 different materials, and in many cases new models are introduced every six months. This means that the environmental issues associated with PCs can change more rapidly than the pace of progress in understanding the impacts. The high-tech nature of the industry also places a priority on proprietary engineering knowledge, so data on process implementation and verification is difficult to obtain. As well, the global nature of the industry—with sourcing and production occurring at many places around the world—poses an additional obstacle when it comes to verifying how production processes are implemented.

The criteria for an eco-label for a PC should address the major environmental issues associated with its life cycle. Chapter 3 discusses three main environmental problems for computer equipment: energy consumption, the impacts of hazardous substances from land-filled computers, and possible exposure to chemicals during production processes. The exposure issue, being entirely in the domain of the implementation of production processes, is difficult to address via eco-labels. The first two issues, however, are more amenable.

Energy - Computers use a fair amount of electricity, with consumption exacerbated by the fact that it is more convenient to leave a machine on 24 hours-a-day.[1] Sleep modes, which automatically put a computer into low-power status when not being used, are very important for reducing electricity use. Also, like many other appliances, computers consume electricity (usually only a few Watts) even when turned off. Eco-labels could (and do) address energy use by setting standards for electricity consumption for different types of computers (laptop versus desktop) and modes of use (full-on, sleep).

---

[1] Electricity use in computers is addressed in detail in Chapter 7.

Hazardous Substances - The main substances of concern that are present in computers are lead, cadmium, mercury, and brominated flame retardant compounds.[2] There are two ways to address hazardous substances via eco-labels. One is to make the absence or reduced content of a material a condition to receive the label. The other approach is to have it ensure "ease of recycling," which affects the feasibility of recycling overall and thus also recovery of sensitive materials at the end-of-life.[3] It should be mentioned though that the actual benefits gained by recyclability depend very much on how collection and recycling is actually implemented. If computers are being sent to landfills anyway, a recycling-friendly design obviously does not do much good.

User-environment issues - Considering the environment at the very local level, around the user, there are a number of issues that could be addressed with eco-labels. These include electromagnetic radiation emissions from cathode ray tube monitors, background noise generated by the computer, monitor performance that minimizes eye fatigue, and ergonomic design of the keyboard and mouse to help prevent carpal tunnel syndrome.

A great variety of eco-labels for PCs have been introduced over the years, beginning with those covering user-environment issues, and then branching out to energy in the mid-1990s, and more recently addressing content of hazardous substances. The main labels are surveyed below and their characteristics are summarized in Appendix 1.

2.1.1 Examples of Existing Eco-labels for PCs

Most existing labels are type I, which means that a third-party organization sets the standards and certifies products under the label. Firms interested in putting the label on their product fill in an application (and usually pay a fee), which is then judged by the issuing organization. In some cases, the application can require documentation from other third-party organizations that test sample products.

1. **Energy Star** – The Energy Star program,[4] launched by the United States Environmental Protection Agency (EPA) in June 1992, is designed to promote energy efficiency via a voluntary, EPA-certified, eco-label on a wide variety of equipment (Figure 1). The range of devices covered by Energy Star has grown significantly over the years; computers and monitors were addressed from the program's inception. For these, Energy Star requires that a device has a low-power sleep mode (one level for computers, two levels for monitors) and sets a maximum for power consumption allowed in that mode (based on the overall size of the power supply). The label has been widely adopted, with around 2,000 desktop models and 3,000 monitors from 130 companies certified. As of 1993, U.S. legislation requires all federal government agencies to purchase only Energy Star-certified computers, monitors, and printers. As

---

[2] The environmental risk posed by the use of these substances in computers was reviewed in Chapter 3.

[3] The recycling of computers is discussed in detail in Chapter 11.

[4] <www.energystar.gov>

well, Japan, Sweden, and New Zealand have adopted the program, with local organizations assuming the task of certification:

- **New Zealand Energy Star** - The Energy Efficiency and Conservation Authority administers the program in New Zealand under the banner of the Energy Efficient Office Equipment Program.
- **Japan Energy Star** - In Japan, it is administered by the Energy Conservation Center.
- **Sweden Energy Star** - The body responsible for the program is NUTEK, the Swedish National Board for Industrial and Technical Development, the central public authority on matters concerning the growth and renewal of the Swedish industry and the long-term development of the energy system.

Figure 1: The international *Energy Star* label.

2. TCO – This is a certification and labeling system developed and managed by the TCO, the Swedish Confederation of Professional Employees, the largest white-collar union in Sweden (Figure 2 and 3).[5] The original focus of the label began with workplace environment issues, such as limiting the electromagnetic radiation from cathode ray tube monitors, sound emitted by devices, ergonomics, and electrical safety. The TCO'95 label also addresses traditional environmental issues, such as energy use, ease of recycling (e.g., labeling of plastic parts), and the content of hazardous substances. The later 1999 version of the certification, TCO'99, addresses cathode ray tube (CRT), flat-panel liquid crystal displays (LCDs), computer systems units, keyboards (traditional and ergonomic), and portable computers. As well, it bans brominated flame retardants in plastic pieces weighing over 25 grams, and places strict limits on cadmium and lead content in batteries. As with the other labels, companies who want TCO'99 on their products support the certification process through a filing fee. It has been widely adopted for CRT and LCD monitors (around 1,700 models certified), and to a much lesser extent, for desktop and laptop computers (19 models certified).

---

[5] <www.tcodevelopment.com>

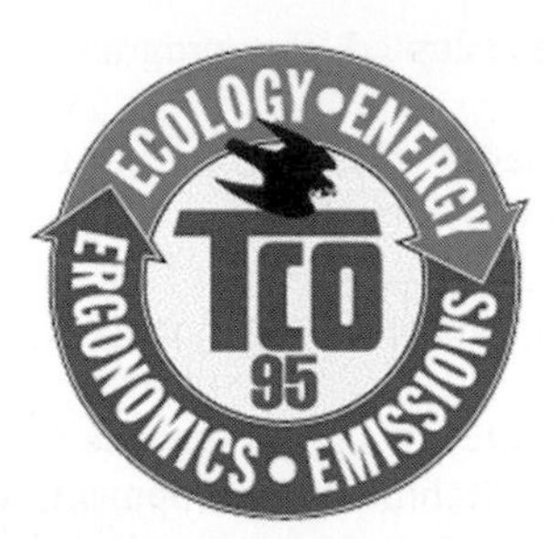

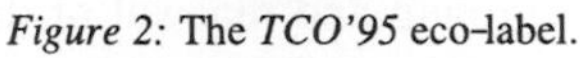
Figure 2: The *TCO'95* eco-label.

Figure 3: The *TCO'99* eco-label.

3. **Blue Angel** – Germany's set of eco-label certifications under the Blue Angel logo[6] was developed by the Umweltbundesamt (German EPA) and covers a wide variety of products (Figure 4). At present, there are around 3,800 products from 710 companies using the Blue Angel label. The requirements for the Blue Angel PC label are similar to those of the TCO'99, but among other things, they also stipulate that no brominated flame retardants be used in circuit boards. Halogen-free circuit boards that meet this requirement are still considerably more expensive than their standard counterparts. Another requirement is that the plastic used in new machines must contain at least 5 percent recycled content, the computer must have at least a two-year warranty, and the manufacturer should carry replacement parts for five years.[7] The Blue Angel PC label has a small market penetration, however. As of June 2002, only three desktops units and three monitors (high-end models) from two firms (Fujitsu Siemens and LG) were certified. The strictness of the certifications, combined with limited geographical coverage of the label, may explain the limited adoption.

Figure 4: Germany's *Blue Angel* eco-label.

4. **Nordic Swan**[8] – This certification label, introduced by the Nordic Council of Ministers, is used in Finland, Iceland, Norway, and Sweden (Figure 5). It too has a PC certification label, similar in most environmental aspects to the TCO'99, but it additionally requires that no chlorinated fluorocarbons were used in manufacturing processes and bans the use of chlorine-based plastics (e.g., PVC). This is a departure

---

[6] <www.blauer-engel.de>

[7] For a more detailed description of requirements, see Chapter 6.

[8] <www.svanen.nu/Eng/>

from other PC eco-labels, which usually only address the characteristics of the product itself, not the production processes. So far, only four models from one (small) Swedish firm have been certified under the Nordic Swan.

5. **E.U. Flower** – Developed by the European Union, the E.U. Flower is an E.U.-wide eco-label program that covers a range of products (Figure 6).[9] The computer label criteria resemble the environmental components of the TCO'99, although the E.U. label also mandates a minimum warranty period of three years. Fees for applications range from U.S.$300 to $1,300, plus an annual fee of $500 to $25,000 (0.15 percent of annual sales of the product in the E.U.). As of March 2003, not one model of computer or monitor has yet been certified. One factor in the lack of industry interest in the label is that the application process reportedly takes two to three months, which can be half the typical six-month product cycle for computers.

Figure 5: The *Nordic Swan* eco-label.

Figure 6: The *E.U. Flower* eco-label.

6. The **Japan Eco Mark** was launched in February 1989 by the Japan Environment Association as a third-party environmental certification program.[10] So far, 64 product categories and 5,176 products the program had been certified (Figure 7), including 84 computers and monitors from six companies (as of June 2002). The criteria of the label for computers and monitors are similar to the others discussed above, plus it bans the use of some brominated flame retardants in circuit boards as well as in plastic casings.

7. Japan's **PC Green Label** – The Japan Electronics and Information Technology Industries Association (JEITA) developed the **PC Green** label (Figure 8),[11] and its criteria cover energy use (must satisfy Energy Star), content of hazardous materials (similar to TCO'99) and ease of recycling issues. It differs from many of the other labels in that JEITA does not certify individual models of computer; rather, companies are given the right to self-declare which of their products meets the standard. While JEITA does not maintain a meta-list of certified products, sources at the association report that some 2,700 computers and 20 firms are using the label.

---

[9] <http://europa.eu.int/comm/environment/ecolabel/>

[10] <www.jeas.or.jp/ecomark/english/>

[11] <http://it.jeita.or.jp/perinfo/pcgreen/>

Figure 7: The *Japan Eco Mark* label.

Figure 8: JEITA's *Green PC* label.

Other countries have also developed their own eco-labels for computers, including Korea, India, and curiously enough, Israel.

2.1.2 Commentary on Eco-labels for PCs

Table 1 summarizes the various criteria for eco-labels. Other than the energy-specific Energy Star, the various labels are roughly similar in the issues they cover, however, the ban of brominated flame retardants in circuit boards, for example, represents a relatively stringent requirement that adds to the price of the final product. No label with this criterion has yet to be widely adopted. As well, the requirement of some labels for a guaranteed two- to three-year warranty is higher than the industry standard of one year.

The adoption of labels is quite varied, with Energy Star, TCO'99, and Japan's Green PC label being the ones used on the widest scale. Looking at the regional distribution of firms applying for different eco-labels, these three are the only ones used by companies around the world, while the other labels seem to be more regional affairs.

There are some aspects of PC production, use, and disposal that are not covered by the existing labels; environmental impacts in the production of equipment and energy consumption in the main use phase (Energy Star only covers sleep mode) are two examples. Some advocates, like the Clean Computer Campaign in California (Clean Computer Campaign 2000), argue that there should also be mechanisms in place to inform consumers of non-environmentally friendly practices (an "anti eco-label" so to speak). This is a valid point, but organizational arrangements to realize such a flow of information is challenging, to say the least. Besides this, details of company practices are confidential and companies are usually not willing to reveal any negative aspects to public scrutiny.

3. CONSUMERS AND GREEN DEMAND: THE USE AND DISPOSAL OF PCS

3.1 Demand for Green PCs

Given the widespread adoption of computers, consumers are now faced with the complex and difficult decision of which PC to buy, with a plethora of models available on the market with varying performance, capabilities, and designs to choose from. Computers are expensive, so cost is an important factor for nearly all consumers. As well, performance and capabilities continue to change rapidly, forcing buyers to keep up with developments in order to make the best choice. Given all these special characteristics of the PC market, the issue of how the environment fits in with consumer choices deserves special consideration.

Publicly available information on consumer attitudes regarding computer purchasing is scarce, thus the authors conducted a survey in Japan of 70 individual users to help understand the consumption patterns and attitudes of consumers towards PCs. One of the survey questions dealt with what issues users considered when purchasing a new computer. Environmental issues were added to the list of factors to get an idea of how receptive consumers might be to eco-labels for PCs, results are shown in Figure 9. The number of consumers listing environmental considerations as a factor was 5.7 percent of all respondents, which, while low in absolute terms, is significant given the general lack of public advertising in Japan of the existence of eco-labels. In retail stores, eco-label logos are displayed fairly prominently on computers; a visit to a large electronics store in the trendy Shibuya area of Tokyo showed that the Energy Star mark was displayed on nearly an equal level as the Microsoft and Pentium logos (Figure 10).

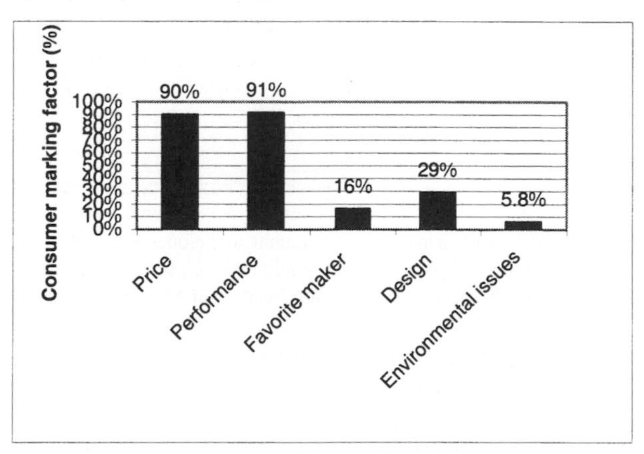

Figure 9: Factors in private consumer's choice of a new PC.

Figure 10: Energy Star labels on computers in Japanese retail stores.

Individual consumers are by no means the only potential target audience for eco-labels: business and government sectors can be as or even more important. In fact, organizations are more amenable to eco-labels than individuals for two reasons: (1) purchasing is done by a specialized group working under organizational guidelines, which, at least in principle, can easily be modified to include environmental considerations; (2) the environmental actions of organizations are far more visible than for individuals, and green purchasing is clearly one way to express environmental commitment.

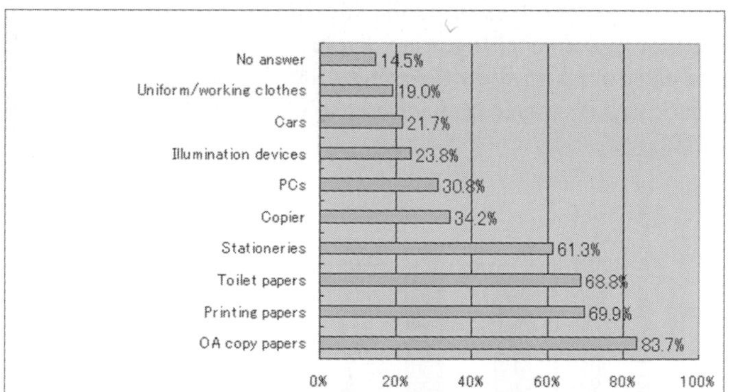

Figure 11: Response of firms and governments to the question, "What kinds of green products has your company purchased for more than a year?"
Source: Nikkei Ecology and the Sanno Institute of Management (1999).

The inclusion of computers in green purchasing policies in Japan has apparently made much progress. A survey of firms and local governments, conducted jointly by the Japanese magazine *Nikkei Ecology* and the Sanno Institute of Management, revealed that 30.8 percent of respondents (out of 506 valid responses) have considered environmental issues when purchasing PCs for at least a year (Figure 11). What "considering" environmental issues means is not precisely defined, but partially reflects whether the organization chose eco-labeled products or not. While substantial, this percentage is smaller than for other categories of goods such as copy paper, which came in with 83.7 percent of respondents reporting that they considered environmental issues when purchasing.

3.2 Green Use of PCs

Consumers have many other environmentally-relevant choices to consider beyond just the decision of which PC to buy. How the computer is used in the home and office also influences the environmental impacts of its life cycle. The decision of *when* to buy a new PC is particularly important, as it determines the effective lifetime of their last computer. Given that much of the environmental impacts of computers occur in their production and then upon disposal (Chapter 3), increasing lifespan is very effective at reducing environmental burdens. Also, how power management functions are used affects the consumption of electricity, an issue further discussed in Chapter 7.

The survey in Japan of individual users also asked how often people purchased a new computer. The histogram of the responses is shown in Figure 12. The average lifetime is 2.7 years. The distribution suggests a significant difference in PC lifetimes, which is not surprising given that computers are used for a variety of activities, and the choice of hardware depends very much on the intended use. As outlined in Chapter 9, it is reasonable to suggest that the users with short computer lifetimes are "power users" who demand state-of-the-art graphics or computing power; while those keeping their computer for two to three years probably use mostly office, e-mail, and Web applications, which have lower computing requirements. This means that there is potential to extend the lifetime of computers owned by power users through resale via secondary markets to "normal" users.

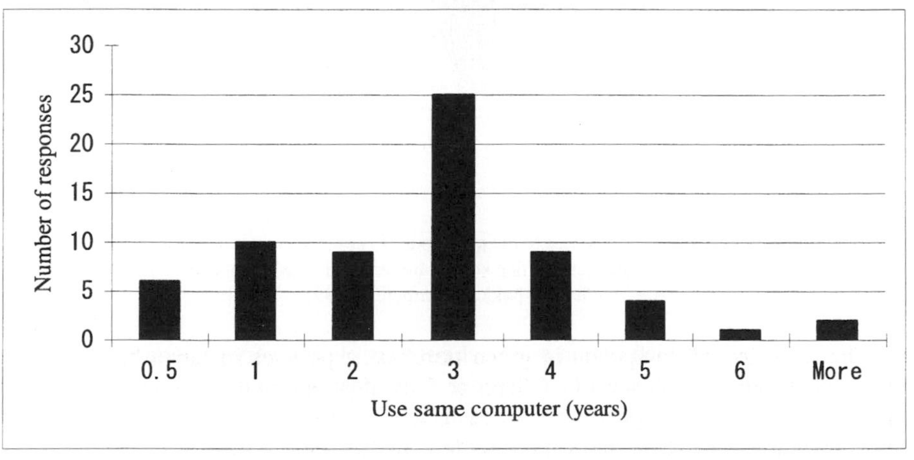

Figure 12: Survey results on the number of years that private users use the same computer.

For Japanese business users, a survey of readers (108 respondents) by *Nikkei Computer* magazine in 2000 found that many companies consider replacing their PCs every three years (Figure 13).

3.3 Green Behavior at the End-of-life of a PC (Green Disposal)

After deciding that it's time to get rid of their "old" computer and buy a new PC, private and business consumers have a number of options available to them. Throwing it away is an obvious choice; around 90 percent of computers in the United States apparently end up in landfills (Amore 1998). Many computer manufacturers have begun accepting used machines for recycling, presuming that the consumer will pay the cost of mailing it to the nearest processing center. It is also possible to sell the computer via a local reseller, classified ad, or Internet auction service, such as eBay. Many donation agencies (as well as friends and acquaintances) also accept used PCs. Quite often, a computer is stored for a number of years until the user eventually chooses one of the options just listed. Many times, people have a strange attachment to their old computer; they're reluctant to part with it for reasons ranging from shock (due to its depreciation in value) to sentimental value. It is often bandied about that 70 to 80 percent of old home and office computers in the United States are stockpiled before any other option is considered (Goldberg 1998; Matthews et al. 1997).

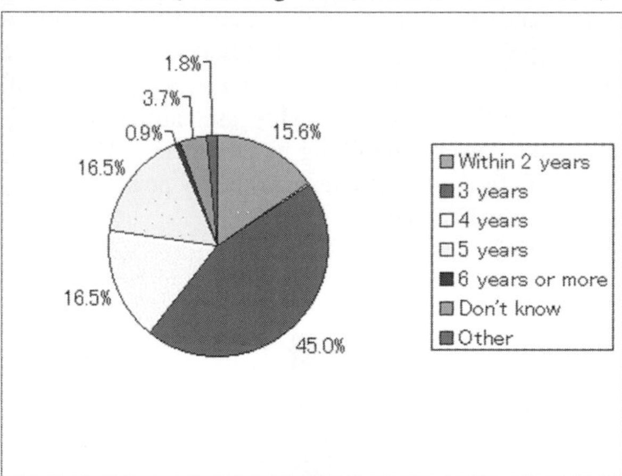

Figure 13: Responses to the question, "In how many years do you think your PC in the office should be replaced with a new one?" *Source*: Nikkei Computer (1999).

The question of which option a user chooses is clearly very important with respect to the environmental impacts of a computer. The ideal option is that they sell or donate their machine as soon as possible. For machines too old to sell or donate, sending them to a recycling center is most preferable. Stockpiling is not desirable from the environmental and economic perspectives, since a PC's re-use value diminishes over time. It also makes recycling (after storing) more difficult as it results in a widening technological spread of computer generations ending up at the recycling centers. The survey of Japanese consumers asked what they did with their computers after they finished using them, given the options to dispose, sell, donate, recycle, or store (results shown in Figure 14). Storage, at least temporarily, is the most popular option, but the 30 percent figure of people storing their computers is much smaller than the numbers in the United States (70 to 80 percent). It could be that the actual amount of stockpiling in the U.S. is lower than believed (no publicly available surveys address this issue), and/or the limited storage space available to the average Japanese consumer presses them to get rid of their old machine sooner. The number of people that said they had sold their used PC is substantial, 23 percent, but this figure may be questionable because the ratio of used versus new machines sold in Japan is estimated to be 7 out of 100 (see Chapter 10). Larger-scale surveys are needed to resolve this issue. The number of users sending their computers to recycling centers is still rather small, but this is not surprising, because many company-run programs have started up only recently and knowledge of the option is not yet common.

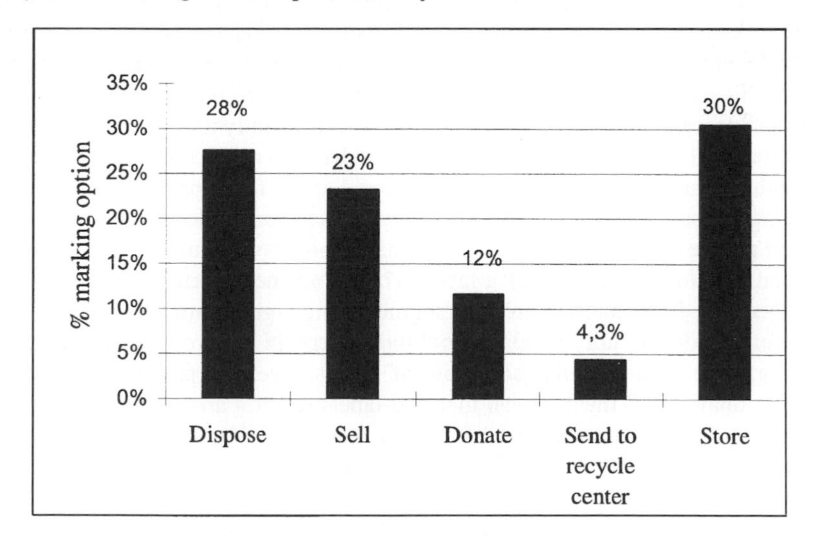

Figure 14: Survey responses from Japanese private users for end-of-life option for PC. (Store = store unused at home)

The survey also asked users whether or not they would donate their old PC to the disabled and the elderly, and if they did and were required to pay, how much would they be willing to contribute. Of the respondents, 80 percent said they were willing to donate, and 61 percent were willing to pay at least something to do so. The average acceptable payment for donating was $12.

4. ANALYSIS AND FUTURE PROSPECTS OF GREEN DEMAND, USE, AND DISPOSAL OF PCS

4.1 Eco-label Systems

Understanding the current and potential contribution of eco-labels for greening PCs is complex. The question can be divided into two related issues. One aspect is how far eco-label standards push computer production and usage towards being more environmentally friendly—or, in more picturesque terms: Is the bar they set high enough? It is sometimes said that eco-label criteria should be set so that only 25 percent of models on the market can receive certification. Such an approach is clearly helpful in general for raising the level of products to that of the "green leaders," but this is obviously not applicable to many of the environmental aspects of computers. Some changes, such as incorporating second-level sleep mode or banning certain substances, are not incremental and may not appear in any model unless stimulated in some way. An eco-label can clearly be a tool to entice firms to make such changes; but they may be effective only when adoption has a minimal effect on manufacturing costs. Given the intensely competitive and international structure of the computer industry, product and process changes that make products significantly more expensive seem out of reach for eco-labels, except for higher-end models where price is not as critical.

The other issue is whether firms actually must apply for approval to use the labels, and whether these certified computers actually replace conventional ones in homes and offices on a large scale. It is often assumed that awareness among private consumers is a pre-condition for the success of a label. This is not necessarily the case, as shown above. Demand from another market segment (e.g., governments, firms) and/or a manufacturer's desire not to appear behind others in environmental performance (marketing) can stimulate the adoption of labels, even if private consumers are completely unaware of them. Given that eco-labels on PCs are not yet widely known by private consumers, the wide use of Energy Star, the TCO, and Green PC labels by manufacturers is probably due to these other factors.

Will the future success of PC eco-labels depend on the knowledge and preferential treatment of them by private consumers? While the answer to this question depends on the issues to be addressed by the label, we argue that, in general, the answer is yes. The reason is that, considering only a special sub-market and environmental reputation, a company may only need to have a small segment of its (higher-end) product line certified under the label. But if most of the products sold are not certified, there is little environmental benefit gained at the macro-level.

How can private consumers be made more aware of PC eco-labels? The general answer is, of course, education, but what sort of education, and how is it to be achieved? The first step is clearly creating the awareness that computers have an environmental impact that ought to be reduced. The popular media has the greatest influence here, and while the issue has received some attention, coverage could clearly be increased. The next step is that consumers learn of the existence and at least something of the significance of the eco-label. In this context it is clearly advantageous that the PC label be part of a larger labeling scheme that includes many other products. The budgets for advertising and raising awareness for many products can thus be combined. But eco-label programs will need to advertise themselves, and some organization needs to take responsibility for producing and financing public awareness-raising activities.

The number of labels put on a product, their presentation, and the information they contain is important. But eco-labels are only one of the many labeling schemes found on PCs; a sample computer for sale in Akihabara in Tokyo sporting 16 different certification logos illustrates the problem (Figure 15). When asked about the labels, the vendors themselves did not know what most of them meant, and there was no explanation of them (including the Energy Star logo) in the machine's documentation. It would be a definite improvement if a brief explanation of the label were provided, including a Web site address, to assist consumers in finding out about the programs their machine has been certified under.

Even if consumers are aware of the significance of eco-labels, it is not obvious how much more they would spend to get a certified computer. Two analyses—the 1996 Globescan Survey performed by Environomics (Canada) and the 1996 Green Gauge Report conducted by Roper Starch Worldwide (U.S.)—studied the willingness of consumers to pay more for green products. These reports indicated that, although the environment is a major concern for the general public, the majority of people need a win-win (environment-economy) scenario to justify a change in their purchasing strategy. In fact, the percentage of Americans willing to pay more for environmental products declined from 11 percent in 1993 to just 5 percent in 1996. One should also note that, in general, consumers still appear to be unaware that the environment is a relevant issue with respect to their choice of computer.

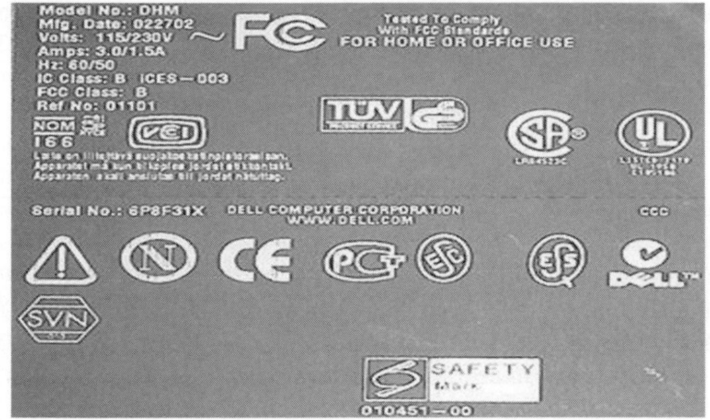

Figure 15: Photo showing the 16 different labels on the back of a desktop PC. Photo taken at Akihabara Electronic Town in Tokyo, December 2002.

It is also important that eco-labels are practicable and recognized by manufacturers. The time factor stands out as being especially important for the computer industry. While a two- to three-month delay for certification might be acceptable for paper products, this poses a real problem for the six-month product cycle of a typical computer manufacturer. The system employed by the JEITA eco-label scheme of certifying a firm to label its own products seems a reasonable solution to this problem. Besides this, a plethora of different regional eco-labels have appeared (many with similar requirements), burdening manufacturers with making many applications and paying considerable fees if they are all to be obtained. The existence of a relatively few, truly international eco-labels is clearly desirable to counter this problem. One possibility is that an international entity, such as the International Organization for Standardization (ISO), the International Telecommunication Union (ITU), or the United Nations Environment Programme (UNEP), becomes the facilitator for a set of global eco-labels. If certain labels could become an international standard, like Energy Star, this would add considerably to their attractiveness for manufacturers.

As an endnote to this section, we mention that (still) none of the eco-labels sets a standard for electricity use in the full "on" mode of computers and monitors. International co-ordination could facilitate an agreement on a measurement standard and an energy-efficiency index to address this. At least for monitors, an energy index recently proposed for TVs could be used as a starting point (Siderius and Harrison 2000). This energy-efficiency index targets a computer's annual consumption, based on the combination of the "on" and "stand-by" modes.

4.2 PC Use

The reason why consumers choose to change computers is important to understand in order to reduce environmental impacts, because this decision determines the lifetime of the machine. While many issues are involved, the desire to upgrade to new versions of operating systems and applications is clearly an important factor. The additional functionality gained from new software, at least for basic functions (i.e., office applications, e-mail, Web browsing), is also apparently diminishing over time, and newer versions of software are often more unstable. It also happens that the hardware demands of new software exceed the improvements, thus leading to a net decrease in speed and convenience. These factors suggest that users could benefit from careful consideration of whether they indeed need a new computer or not. As consumers become more aware of the environmental impacts of PCs, the most environmentally friendly action they can take is simply to delay purchasing a new machine, which is also good for their pocketbook (presumably they do not use the money they save to help buy an SUV!).

4.3 End-of-life

There is clearly untapped potential for end-of-life computers to bypass storage in the closet and more efficiently move to used and donated markets or recycling centers. Users derive no benefit from stockpiling and can potentially generate income from the sale of their used PC. The survey results on the distribution of computer lifetimes and disposal options suggest that there may be significant potential for expansion of the market for used PCs. A significant number of "power users" buy a new machine after only six months or a year, which could easily serve well the needs of those users only needing office applications, Web browsing, and e-mail; yet many users continue to store their old, unused machines. The reasons for this need to be further explored in order to clarify how to improve the timely flow of machines to used markets and recycling centers. Concern for data security is often cited as a reason PCs are stored; improved knowledge of proper deletion of personal data would help ease such fears. Many users may also be unaware of where and how to sell their old computer. Also, knowledge of how to send the computer to the manufacturer's recycling center needs to be further disseminated.

5. CONCLUSIONS

Consumers affect the environmental performance of their computers through their decisions along the entire life cycle. These decisions include which computer to purchase (eco-labeled or not, which eco-label to look for), how to use power management functions, how long to keep a computer, and what to do with a PC at its end-of-life. While some of these choices could mean that they spend more money (e.g., choosing a computer certified by an eco-label with particularly tough standards), many, if not most, choices are economically neutral or even positive for consumers. Yet it seems that the "rational" choice is often not made. While a number of factors could explain this, awareness of the rational course is an obvious pre-condition to choosing it. There is clearly still a lack of information available to consumers regarding their purchase, use, and disposal of computers; thus a priority should be set on dissemination activities. One preliminary step is to increase the awareness of consumers of the fact that their computers have an impact on the environment; without this base, they have no reason to consider the issue at all. The popular media is probably the best way to raise awareness, but increased and regular coverage of the topic is necessary. At the next level there is a need to make available more detailed information on what actions are environmentally preferable. One obvious step that could be taken is for the documentation that comes with new computers to include a brief "environmentally friendly user" guide that explains the eco-labels the machine is certified under, how to use power management features, how to upgrade the computer, and where to send it for recycling. The popular media can also play a greater role in explaining to consumers their options of selling or donating their used PCs.

There is a whole level above these simple steps that also needs consideration in the future. For one thing, the plethora of different PC eco-labels should probably be consolidated under several standard international labels that cover different groups of issues. This would lend credibility, improve comprehension, and also ease the burden of manufacturers in the application process.

There are other issues to tackle here as well, but it is evident from the discussion above that there are a number of clear and promising directions to pursue—from the perspective of consumer choices—to reduce the environmental impacts of PCs. The next step is for the relevant actors to continue efforts to plan and implement the next stage.

ACKNOWLEDGEMENTS

We are grateful to a number of individuals that have helped carry this project to this stage. We are indebted to Eric Williams for editorial assistance with the various revisions of this chapter; his comments and suggestions were very helpful in finalizing it. We further express our gratitude particularly to Eiji Miyazaki of the University of Electro-Communications in Tokyo and Gareth Johnston of the United Nations University Press for their invaluable assistance.

Appendix 1: Summary of PC eco-labels.

Label	Date Criteria Adopted	Energy Consumption							Safety				Re-use, upgradibility	Recycle		Certified products by end of 2002	
		Desktop system Unit		Monitor CRT/LCD		Portable computer			Noise limit	Electro-magnetic emission limit	Chemical bans						
		Sleep mode (W)	Off mode covered	Sleep mode (W)	Deep sleep mode	Sleep mode (W)	Deep sleep mode	Off mode			Lead, cadmium, mercury	Flame retardants		Product take-back	Easy re-cycling	No. companies	No. models certified
PC Green Label (Japan)	1 Jul 2002	<15	no	<8	no	<15	no	no	yes	yes	case, battery *1	case *1	yes	yes	yes	20	2700
Eco Label (Japan)	1 Dec 2000	<15	no	<15	<8	<15	no	no	48 db ISO9296	VCCI *2	case, battery	case	yes	yes	yes	6	84
Energy Star (USA)	1 Jul 1999	<15, *3	No	<15	<8, After 60'	<15, *2	no	no	no	no	no	no	no	no	no	103	2412
EU EcoLabel	22 Aug 2001	<27	yes, <5	<10	<3	<5	<3	<2	48 db idle, 55 db operating	low freq. magnetic fields <200 nT	*4	*1	yes	yes	yes	0	0
Nordic Swan Ecolabel	15 Mar 2002	<30	no	<15	<8	<8	no	no	48 db idle, 55 db operating, ISO9296	CE labelling	case, battery *4 *5	case, circuit board	yes	yes	yes	4	1
Blue Angel (Germany)	# *6	<30	no	<10	<5	<7	<5	<2	LWAd 55 dB(A) *7	TCO99 for magnetic fields.	case, battery	case, circuit board	yes	Yes	yes	2	6

(Cont.)

(Appendix 1 cont.)

TCO 99 (Sweden)	31 Dec 1999	<30	<8	<10	<5	#	#	#	48 db idle, 55 db operating	low freq. magnetic fields <200 nT	case, battery	case	yes	no	yes	127	1746 dis-plays, 3 notebooks, 18 system units, 1 keyboard
Korean EcoLabel	#	<27	<5	<10	<10	#	#	#	<50 db	no	all	#	no	yes	yes	1	2
Taiwan Ecolabel	#	<30	no	<10	<5	#	#	#	#	#	no mercury battery	#	#	#	#	#	#

Notes:

*1. Lead, cadmium, mercury and specific bromide flame retardants (PBBs, PBDEs) must not be contained in plastic frames weighing more than 25 g (i.e., PC enclosures, monitors, keyboards, mouses and batteries).

*2. VCCI: Voluntary Control Council for Interference by Information Technology Equipment.

*3. The Energy Star was previously <30, but after July 2000 <15 (after 30 minutes).

*4. A maximum average level per lamp is 3 mg in LCD monitors.

*5. Cadmium and lead are not applied to plastic parts weighing more than 25 g.
Content of mercury, cadmium and lead in batteries and accumulators must not exceed 5 ppm for cadmium (Cd), 5 ppm for mercury (Hg) and 15 ppm for lead (Pb).

*6. # symbol indicates no information available.

*7. LWAd is a measure of sound level known as "A weighted."

Sources:

Energy Star <http://www.energystar.gov/> and <http://www.energystar.gov/ia/products/ofc_equip/purchtips.pdf>
TCO <www.tcodevelopment.com/>
Blue Angel <www.blauer-engel.de/>
EU Flower <http://europa.eu.int/comm/environment/ecolabel/>
Nordic Swan <www.svanen.nu/Eng/>
EcoMark <www.jeas.or.jp/ecomark/english/>
PC Green Label <http://it.jeita.or.jp/perinfo/pcgreen/>
Taiwan Ecolabel <http://www.epa.gov.tw/english/LAWS/ecolable.htm#1.Translators'%20Note:>

REFERENCES

Amore, D. 1998. Study finds computer recycling not clicking. *Waste Age* 30(12): 14–15.

Clean Computer Campaign. 2000. *Right to know a little...Exposing double standards in global high-tech production.* San Jose, CA: Silicon Valley Toxics Coalition. <http://www.svtc.org/cleancc/pubs/2000 report.htm> (19 December 2000).

Goldberg, Carey. 1998. Where do computers go when they die? *The New York Times*, 12 March. <http://gdi.ce.cmu.edu/comprec/nytimes98/12die.html> (14 May 2003).

Matthews, H. Scott, Francis McMichael, Chris Hendrickson and Deanna Hart. 1997. Disposition and end of life options for personal computers. *Green Design Initiative,* technical report #97–10. Pittsburgh: Carnegie Mellon University. <http://www.ce.cmu.edu/GreenDesign/comprec/NEWREPORT.PDF> (14 May 2003).

Nikkei Computer. 1999. *Markets in Japan.* Tokyo: Nikkei Business Publications. <http://www.nikkeibp .com/markets/archives/index991126.html> (14 May 2003).

Nikkei Ecology and the Sanno Institute of Management. 1999. *Markets in Japan.* Tokyo: Nikkei Business Publications. <http://www.nikkeibp.com/markets/archives/index001128.html> (14 May 2003).

Siderius, Hans-Paul and Robert C. Harrison. 2000. An energy efficiency index for TVs. Proceedings of the Second International Conference on Energy Efficiency in Household Appliances and Lighting, 27-29 September at Naples.

Young, John. 1994. *Global network: Computers in a sustainable society.* Washington, D.C.: World Watch Institute.

Chapter 9

STRATEGIZING THE END-OF-LIFE HANDLING OF PERSONAL COMPUTERS: RESELL, UPGRADE, RECYCLE

Eric Williams[a] and Yukihiro Sasaki[b]
[a]*United Nations University, Japan*
[b]*Mizuho Financial Group, Japan*

1. INTRODUCTION

The question of how to deal with personal computers (PCs) and other information technology (IT) equipment when they have reached the end of their useful lives is increasingly on the minds of those in governments, industry, and the public.[1] Much of the discussions and activities have focused on how PCs may be most efficiently collected and recycled. While recycling is clearly necessary, the traditional wisdom of waste management dictates that upstream management of wastes is as important, if not more important, than final treatment. Upstream management refers to strategies to reduce the size of the incoming waste stream by fully using and reusing products *before* they are thrown away. The idea of emphasizing these aspects has already been codified in waste management concepts such as the "3Rs" strategy—reduce, re-use, and recycle.

Such strategies should also apply to IT equipment; indeed, one can argue that there is even more potential in emphasizing upstream management with IT than for many other products. One reason is that most computers disposed of are in perfect working order; in most cases the original user switches to a new machine to gain the benefit of more recent technology, not because of physical wear on the computer. The other is that, compared to other manufactured products, the major impacts of producing

---

[1] As some of this analysis is fairly technical in nature, a non-technical summary of the entire chapter appears at the end after Section 5.

Computers and the Environment: Understanding and Managing Their Impacts
Edited by Ruediger Kuehr and Eric Williams, pages 183–196.

computers occur during the processing of materials and manufacture of specialized parts, as opposed to the production of the raw materials that go into them (see Chapter 3). The result is that recycling raw materials does not "refund" environmental impacts nearly to the extent that it does for simpler products such as aluminum cans and newspapers.

Upstream management of the computer waste stream is essentially about more fully meeting the needs of users with existing machines, resulting in the production and disposal of fewer PCs. Extending a computer's lifespan can be done in three ways: delaying the purchase of a new machine as it is, reselling to secondary markets, or upgrading. While the various actors dealing with waste electronics are certainly aware of these options, they have yet to be explicitly analyzed and aggressively pursued. In the European Union, recently passed legislation dealing with waste electronics (WEEE, RoHS), for example, focuses on recycling and trying to keep hazardous materials out of computers in the first place (European Parliament 2003a, 2003b). Eco-labels for computers address PC lifetime issues by ensuring extended warranties and supplies of spare parts (see Chapter 8). The decision to purchase a new computer, however, is almost never about the breakdown of the old machine; instead it is driven by improved capabilities of the newer machines.

The argument that reselling and upgrading are environmentally friendly has thus far been theoretical. Recommending policies or other forms of social response should be supported by analysis that endeavors, to the extent feasible, to verify what actions are indeed appropriate. To this end, we have undertaken a quantitative analysis of the environmental benefits of reselling, upgrading, and recycling PCs in terms of life cycle energy use. From the start it is clear that uncertainty will remain in the final result (mainly due to data issues) and that such analysis does not describe the whole picture. It is, however, a useful and do-able starting point. At the outset it is worth emphasizing that energy use is not the only relevant environmental issue for computers, but when comparing re-use options, all environmental burdens become linked to a large degree. A savings in energy is indicative of production (and disposal) of fewer new units, which in turn implies reduction in environmental burdens across the board.

2. RESELL, UPGRADE, AND RECYCLE OPTIONS

Before launching into the actual energy analysis, we first explore the different end-of-life options and their respective states of implementation.

2.1 Resell

Reselling is the practice of selling a computer to a secondary consumer when it reaches the end of its useful life for the original purchaser. In this case, a zero price (donation) of a computer for some social purpose, such as education, is included in the definition of reselling. Secondary markets for computers are distinct in character from those of other products, such as automobiles or furnishings, as computer markets are driven by the depreciation of quality relative to newer models, as opposed to decreasing absolute quality or reliability. Used-PC markets are discussed in detail in Chapter 10.

2.2 Upgrade

Upgrading a computer refers to the replacement of certain components with newer versions in order to improve performance. Often, the main motivation is not dissatisfaction with the computer using the original software, but rather the desire to keep pace with the increasing demands of new operating systems and applications. The word *refurbishing* is also used, but this term is avoided here as it connotes restoration to original condition, while for computers the goal is to improve the machine's capabilities. A typical upgrade usually involves replacing the microprocessor, memory and/or hard drive—though more extensive upgrades are possible. A typical upgrade is considerably less expensive than purchasing a new PC. For example, a 2-GHz Pentium IV processor, 128-MB RAM addition, and a 20-GB hard drive were priced at U.S.$190, $30, and $80, respectively (February 2003). Not all capabilities, however, can be easily upgraded. For instance, bus speed requires an entirely new motherboard—a fairly expensive proposition. Also, the introduction of different types of ports to interface with video (e.g., IEEE 1394) and/or peripheral buses (e.g., USB) can result in a "complete" upgrade costing more than a new machine.

The extent to which upgrading is currently practiced is undocumented. Interviews with companies in Tokyo that specialize in computer services suggest that less than a few percent of users opt for an upgrade, and that all their clients are computer hobbyists/specialists to some degree. These firms emphasized the necessity of user knowledge in upgrading a computer. Although an upgrade may be economically advantageous, a relatively small fraction of users are sufficiently informed to recognize this fact. As society becomes more literate in information technology, this fraction will likely increase.

2.3 Recycle

Recycling involves disassembly and/or destruction of a computer in order to recover parts and materials. The term *demanufacturing* is also used, which connotes an emphasis on re-use of parts over "liquidation" to recover raw materials. Recoverable raw materials fall into the categories of metals, glass, and plastics.

Metals make up about half the weight of a typical desktop PC. The technology for recycling metals is fairly well developed, and existing facilities can recover steel, aluminum, copper, nickel, lead, zinc, gold, silver, and platinum from waste computers. The latter three precious metals are mostly used in printed circuit boards, which also contain environmentally significant quantities of hazardous metals, such as lead, mercury, chromium, and cadmium.

The glass used in cathode ray tubes (CRTs), which makes up 28 percent of the weight of a typical desktop system, is difficult to recycle due to the need for deconstruction and separation of components. Typically, CRT glass is divided into four categories according to lead content and then shipped to glassmakers, such as Corning Asahi, for use as raw material for specialized products (Dillon 1998)

Plastics make up 23 percent of the weight of a computer system, but many technological and design barriers still remain with respect to recovery. The main obstacles are associated with the different mixture qualities of the plastics (Das 1999). A large variety of plastics are used, and there are few processes available to separate these into useable sub-components. Techniques using a blast furnace have apparently been successful in separating out re-usable polycarbonate (PC), polystyrene (PS), polyethylene (PE), and acrylonitrile-butadiene-styrene (ABS) (Takesue 2000). Another approach is to design PCs using fewer kinds of plastics and make them easier to disassemble, so as to facilitate recycling.

The main components in a computer that can be re-used include the fans, transformers, wiring, and disk drives. There is little data available indicating to what extent these can actually be used. One report on the economic breakdown of income from scrap electronic processing suggests that fans, transformers, wire, and disk drives make up 8 percent of the weight of electronic waste and contribute 11 percent of income (Pepi 1998). The re-use rate for microchips and other components on a circuit board is unknown. Circuit boards command the highest selling price among electronics wastes (about $1 per kilogram), but this is apparently due to the value of the precious metals they contain rather than the components.

A study by the U.S. National Safety Council reports that 6 to 11 percent of PCs are recycled, indicating that it has yet to take off in the United States (Amore 1998). The level of computer recycling in the European Union and countries such as Japan will dramatically increase as various legislation mandating the take-back of electronics comes into effect.[2]

3. ENERGY USE IN DIFFERENT PHASES OF THE PC LIFE CYCLE

The first stage of the analysis is to build a picture of energy use in different stages of the computer life cycle—production, resale, use, upgrade, and recycle. The case considered here is a desktop computer with a 17-inch CRT monitor used at home. Results will vary somewhat for laptops, LCD monitors, and office use patterns—an issue to be discussed in the sub-section on caveats below in Section 4.

[2] Computer recycling technology and practices are discussed further in Chapter 11.

3.1 Production

Estimating the total energy it takes to make a computer requires adding up the contributions from the various industrial activities along the production chain (i.e., semiconductors, circuit boards, disk drives, specialty chemicals, and assembly). This kind of analysis is the domain of life cycle assessment (LCA) (Curran 1996). Chapter 3 of this volume included the first estimate of energy used in producing a computer to fully disclose underlying data and assumptions. The result was that some 5,040 megajoules (MJ)[3] of energy are required to produce a desktop computer with monitor. An alternative methodology known as economic input-output life cycle assessment gives a result of 5,600 MJ per computer (Hendrickson et al. 1998) (see also Chapter 3, Section 3.6.2). The latter figure is used, as it includes additional industrial activities such as making the equipment to produce computer components.

3.2 Use

Electricity consumption during the use phase requires information on usage patterns, power consumption in different modes, and the lifetime of the device. A typical desktop unit with a CRT monitor uses 115 Watts of power in the active mode (Miyamoto et al. 1998). Given the lack of publicly available studies on usage patterns, we assume a scenario of three hours of use per day, 365 days per year. This is likely an overestimate of the active mode use for home users, but, hopefully, accounts for power consumption in standby (sleep) mode.

The survey in Japan of 70 users (Chapter 8) suggests an average computer lifespan of 2.7 years. Another recent survey in Japan of 1,350 Web-users reports that they purchase a new computer every two years on average (Lifestyle Center 2002). Two- and three-year lifespans will be considered in this analysis. The result for use-phase consumption is 910 MJ (250 kWh) for a two-year lifespan and 1,370 MJ (380 kWh) used in a three-year lifespan.

3.3 Resell

The central questions regarding resold PCs are the extent to which they replace the purchase of new machines and also the length of the second lifespan. In the absence of publicly available data on this point, we assume that the purchase of a used PC does indeed substitute for new demand and that the second lifespan can be either one or two years. The latter is based on a survey of 17 used-PC vendors in the Akihabara electronics store district in Tokyo. While these figures seem reasonable, they are based on the impressions of vendors, who do not usually follow up on what consumers do

[3] One megajoule (MJ) = 1 million joules, a standard measure of energy in the metric system. Some equivalent measures are 1 MJ = 239 kilocalories (kcal) = 948 British thermal units (BTU) = 0.28 kilowatt-hours (kWh). One liter of gasoline contains around 35 MJ of heat energy.

after purchasing the used PC. The actual length of time that purchasers continue using their second-hand PC should be analyzed via surveys of actual behavior—a task beyond the scope of this analysis.

3.4 Upgrade

It is reasonable to assume that upgrading replaces the purchase of a new PC. In most cases it extends the lifetime of the same user's machine, assumed to be either one or two years. As with reselling, this range is based on surveys, in this case, the companies in Akihabara that supply upgrade parts and services.

For upgrading there is also the issue of the energy required to produce the component parts. This energy will be estimated by an economic input-output LCA. As mentioned in the previous section, the cost of a new CPU and 128 MB of RAM is $220, while a new hard disk runs about $80. The energy intensities of the semiconductor and related devices (computer peripherals) are 5.6 MJ per dollar and 6.7 MJ per dollar, respectively (Green Design Initiative 2003). The result is that the energy embodied in the parts for an upgrade is 1,750 MJ. This is a significant fraction of the total production energy (5,600 MJ), essentially because semiconductor production consumes considerable energy despite the negligible physical weight of the chips themselves (Williams et al. 2002).

3.5 Recycle

It is very difficult to estimate the energy balance of computer recycling due to the lack of publicly available data on recycling processes. Because of this, we are forced to resort to using a "black box" result (i.e., assumptions and data used are largely unreported). A study done by computer manufacturer NEC on their own state-of-the-art system to recycle desktop PCs reports a reduction of 33 kilograms (kg) of carbon dioxide (CO_2) emissions, equivalent to 280 MJ of energy, compared to disposing of the machine in a landfill (Takesue 2000). The analysis reportedly includes transport, disassembly, and processes for recovering plastics, metals, and glass, plus producing the virgin materials that the recycled ones replace.

We also consider the possible lower and upper bounds of the energy credit or cost for recycling computer materials. For the lower bound, note that the net economics of recycling computers is negative—reputedly costing $10 to $30 per machine. This suggests that the net energy balance could also be negative; i.e., more is expended on transport and processing waste IT equipment than is recovered from the materials recycled. This is probably not the case, because dismantling processes are usually less energy-intensive than the materials production sectors they replace; but there is no publicly available evidence to justify discarding a net energy cost for recycling computers as a worst-case scenario.

It is also worth estimating the ideal upper-limit of the energy credit from recycling the materials from used computers. This is done by assuming that transport and processing of the computer has a zero energy cost and that all materials can be 100-percent recycled. The energy embodied in the raw materials can be estimated by combining a total of materials with their production energies. This analysis was done in Chapter 3 (Table 6 and 7), with the result that 1,570 MJ of energy are needed to produce the raw materials for a computer, about 28 percent of total production energy. It is physically impossible for the energy credit from recycling raw materials to exceed this amount.

3.6 Transport

Energy is also expended during the transport of computers and parts involved in reselling, upgrading, and recycling. In the process of reselling, users may bring the computer by automobile to a local reseller or ship it via courier service to another user or large reseller. For upgrading, the user may drive to a local shop to buy the parts or may order online for delivery via courier service. With recycling, because the number of computer recycling centers is still relatively small, the distance between the consumer's home and the destination can be in the hundreds or even thousands of kilometers.

To first estimate how large the energy use of transport could be, one can calculate that a 10-kilometer round-trip via personal vehicle uses around 33 MJ. This figure assumes a vehicle mileage of 13 kilometers per liter and includes production of the car and the use of gasoline (Williams et al. 2003). In Japan, macro-statistics suggest that shipping a 24-kilogram (kg) computer 1,000 kilometers via truck requires around 136 MJ of energy; this includes production of the truck and production and use of the diesel fuel (ibid.). The transport energy for a worst-case scenario of shipping the computer 5,000 km (e.g., New York to Los Angeles) is thus 680 MJ. This is around 10 percent of the energy used in the production of a computer; thus transport can affect the final result to some degree.

It is, however, very difficult to accurately estimate how much energy is needed "on average" for transport. For instance, how often do consumers sell their PC locally, versus on an Internet auction site like eBay? There are no statistics currently available to deal with the many unanswered questions. Still, it is not difficult to estimate a plausible order of magnitude for typical and worst-case scenarios; this should be sufficient, as the transport factor does not seem so important in resale and upgrade options. For recycling, energy use for transport is of a similar size to plausible recycling gains, and thus could significantly affect the outcome.

The base transport case for reselling a computer will assume that the seller and buyer both drive to and from a local shop on a ten-kilometer, single-purpose trip (e.g., no other shopping done on the trip). The worst case shipping distance is assumed to be 5,000 kilometers. The estimated low-end and high-end of transport energies for reselling are thus approximately 70 MJ and 700 MJ, respectively.

Transporting a computer for upgrading assumes that the user orders a two-kilogram package of parts (processor, memory, hard drive) via the Internet from a vendor 3,000 kilometers away. Alternatively, the user makes a ten-kilometer, single-purpose trip to and from a local store. In either case, the energy consumed is around 30 MJ.

3.7 Summarizing Energy Results

The values for energy use and lifespan are summarized in Table 1, along with the definition of notation to be used in the next section. Note that landfilling a computer is not included in the list; apparently its energy cost is negligible compared to other factors (Miyamoto et al. 1998).

Table 1: Energy and other parameter values for a desktop computer (home use).

Life cycle stage	Notation	Value
Production	E_M	5,600 MJ
Electricity use:	E_{Use}	
2-year first lifespan (base case)		910 MJ
3-year first lifespan		1,360 MJ
Second lifespan		
base case		2 years
worst case		1 year
Upgrade (energy to make parts)	E_{Up}	1,750 MJ
Recycle:	E_R	
Base case		–280 MJ
Idealized best case		–1,570 MJ
Transport of used PC/parts:	E_T	
20 km car trips for PC (base case)		70 MJ
5,000 km shipping of PC (worst case)		700 MJ
3,000 km shipping of parts (base case)		30 MJ

4. EVALUATING THE ENVIRONMENTAL EFFECTIVENESS OF RESELLING, UPGRADING, AND RECYCLING

4.1 Model Definition

The basis of the analysis is a simple mathematical model that represents the relationship between life cycle energy use and the degree of implementation of end-of-life options. The life cycle energy (LCE) associated with a set of computers can be modeled as:

Equation 1

$$LCE = E_M - x\, r_1(E_M - E_T) - x\, r_2(E_M - E_{Up} - E_T) + E_{Use} + r_3 E_R$$

In this equation, r_1, r_2, and r_3 are the implementation rates (percent) of reselling, upgrading, and recycling for a set of computers, and x is the ratio of second lifespan over first lifespan for resell and upgrade options. It is considered the same for both reselling and upgrading (though in practice it will probably be different). x is valued at 1 when both first and second lifetimes are two years, or 1/3 when the first lifetime is three years and the second lifetime is one year. The central assumption in the model is that reselling or upgrading a computer will replace the need for a new machine for the span of its second life, after which that user gets a new machine. This model also assumes no interaction between the rates of implementation of different options. This is not true in general but should hold when implementation rates are small.

The effectiveness of end-of-life options in saving energy can be quantitatively modeled using the above formula. The effectiveness coefficient (EC) is defined to be the percentage savings in life cycle energy given 10 percent implementation of a given end-of-life option. In mathematical terms, the effectiveness coefficient of option j (where 1 = resell, 2 = upgrade, 3 = recycle) becomes:

Equation 2

$$EC_j \equiv \frac{\Delta LCE}{LCE} = \frac{\frac{\partial LCE}{\partial r_j} \cdot 10\%}{LCE}$$

4.2 Results

Numerical values of the effectiveness coefficients are calculated using the values in Table 1, with the results shown in Table 2.

Table 2: Fraction of life cycle energy saved, given that 10 percent of computers are resold, upgraded, or recycled.

Scenario	EC_1 (resell)	EC_2 (upgrade)	EC_3 (recycle)
Base case	8.6%	5.2%	0.43%
Long 1st life (3 yr), short 2nd life (1 yr)	2.7%	1.8%	0.40%
Worst case transport scenario (used PC shipped 5,000 km)	7.5%	-	-
Recycling at theoretical best (E_R=–1,570 MJ)	-	-	2.4%

For the base case, reselling a computer is some 20 times more effective at saving life cycle energy use than recycling and upgrading is 12 times more effective. Other scale-driven environmental impacts should also show a similar gap. The results show a dramatic reduction in the benefits of reselling and upgrading as the second lifespan gets smaller, though these still save more energy than recycling, even given pessimistic assumptions. The conclusion to be drawn here is that the emphasis given to upstream waste management according to the "3Rs" should also be applied to computers.

The results can also be understood from an intuitive perspective. Since computers are high-tech products, it is natural to expect that the bulk of the production energy is embodied in the form of the product rather than in the raw materials. As the rapid technological cycles reduce the reusability of the parts themselves, recycling becomes primarily focused on the raw materials; but as most of the energy investment is in the form of the product, not its materials content, the extension of life span naturally leads to greater savings than recycling.

4.3 Caveats

If numbers were always true, this section would not be necessary; but outside the realm of pure mathematics, analyzing real world problems always requires making assumptions, sometimes "heroic" (or "villainous," depending on one's point of view), and depends on outside data of often unknown quality. The results of such analyses can be wrong, and this danger should always be carefully considered. Many studies, unfortunately, do not and indeed often neglect to report data and assumptions, so outsiders can't form an opinion about the accuracy of the study. While we have certainly labored to detail the basis of the analysis, it is also appropriate to highlight what we see as its potential weaknesses.

In the present case, the discussion of caveats can be divided into issues of data quality, scenarios considered, and issues related to the underlying model used. Data quality is clearly important, but much uncertainty remains regarding energy use in different stages of the computer life cycle. The figures used for energy in production and use are probably correct within some tens of percent. Recycling is quite a different story; we dealt with a large uncertainty by also considering the energy payback for idealized "perfect" recycling to put some limits on the possible range. Results are also sensitive to values of the first and second lifetimes. The values used for the latter are taken from vendor's impressions—they should actually be based on detailed surveys of user behavior. The uncertainties in the second lifetime were handled by also considering how the results change under a pessimistic assumption of its value being only one year. In terms of scenarios considered, the analysis treated only the case of desktop computers for home use. Admittedly, there are many office users (for which use phase consumption is much higher), plus laptop computers are increasing in popularity (lower use-phase electricity consumption). We also analyzed office users and laptop cases, where the figures for energy savings change of course, but the overall pattern (large resell and upgrade savings compared to recycling) is the same as for desktop home users. With regard to the model itself, it assumes no relation between the implementation of resale, upgrade, and recycling. If the different options are implemented on a large scale, they will interact in practice (e.g., resale market affects scale and age of machines entering recycling processes).

These caveats should, however, be considered in the context of the objective of the analysis. The purpose here is not to compare end-of-life options to an accuracy of several digits, but simply to understand the order of magnitude. From this perspective, the results should be qualitatively robust, and it is fair to assert that reselling and upgrading are very effective options for reducing life cycle impacts of computers.

5. POLICY IMPLICATIONS

The central task of this chapter has been to build a quantitative case showing that reselling and upgrading of computers is particularly effective at reducing environmental impacts compared to recycling. Even though much work remains to improve the accuracy and depth of the result, we argue that it should already be considered in the formulation of policies relating to waste electronics. Policymaking must always cope with uncertainty. If we always waited for complete knowledge, nothing in business, government, or private life would ever be done. The main principle to guide us through uncertainty is to base decisions on the best available knowledge. Since this analysis, combined with that in Chapter 3, represents the first attempt to both evaluate end-of-life options, as well as disclose all data and assumptions, we believe it deserves consideration by policymakers.

What are the policy implications? To comment on this, the discussion needs to be expanded beyond environmental issues. Economics, user needs, and their knowledge play important roles both in determining the potential of reselling and upgrading computers, and also in how policy could stimulate them.

The macro-economic implications of increased reselling and upgrading are not obvious. More reselling and upgrading implies reduced sales of new computers, but means expansion for reselling and upgrading service sectors. This shift will hurt firms focusing on selling new machines; but many manufacturers are increasingly expanding their business areas to include services. Firms such as Dell and IBM, for example, sell used machines via their Web sites and also lease computers. Consumers save money by reducing purchases of new machines, which will be saved or spent on other goods. This is positive from the point of view of the individual consumer, but will this redistribution of spending be "good" for the overall economy? The answer is not obvious; there are arguments on both sides and the answer may well depend on local conditions. It is perhaps fair to say that, while there are economic winners and losers in the improved use/resell/upgrade scenario, the overall outcome for the economy is difficult to predict, and thus should, for the time being, be assumed as neutral.

The economic costs of recycling computers should also be taken into consideration. For example, recently enacted legislation in Japan mandates the recycling of computers by private and business users. The system is funded via non-visible fees paid by the consumer at the time of purchase. The fee is expected to range from $20 to 40 per computer (*The Japan Times* 2002), and given the ten-million-unit annual market for new computers (JEIDA 2000), the total cost of the recycling system is some $200 million to $400 million per year. This is a substantial sum, and promoting increased reselling and upgrading could be inexpensive options to mitigate the overall cost of the system.

The computing needs of users and their knowledge of them also play an important role. It would not be practical to ask consumers to buy used computers or upgrade their old ones if everyone truly wanted (and is able) to pay for a new computer every two years or so. The vibrant market for used PCs in the United States is an indication that there is significant demand for them. As discussed in Chapter 10, a number of "non-economic" obstacles remain to full realization of this market, including difficulties in transferring licenses for pre-installed software and the tendency for users to store a computer even after purchasing a new one. The former issue can easily be addressed via legislation (or actions by companies) to ensure the smooth transfer of licenses. The latter relates to whether users know how to take advantage of the used-PC market. Public awareness campaigns and media coverage are two tools available to raise awareness. The potential of upgrading and the means to encourage it are less clear. Designing computers to be easily upgradeable is clearly helpful, but just because a machine *can* be upgraded does not mean that the user *will* choose to do so. The lack of knowledge of how and why to upgrade may be an obstacle. Companies assisting customers in upgrading report that, in Japan at least, all clients are computer hobbyists or specialists to some degree. This client base will likely increase as society as a whole becomes more familiar with information technology, but will this represent a significant fraction in the near future? One possibility is for manufacturers to offer information on upgrade packages to existing users every year or so. While this may (or may not) have business potential, many manufacturers may be reluctant to allot attention away from what they perceive as their main focus—selling new machines. One possibility is that offering upgrade services could be made a requirement of computers eco-labels (see Chapter 8).

The above discussion reflects a few ideas for improving the degree of reselling and upgrading. There are many other avenues to pursue, and appropriate actions could well vary according to national and regional conditions. What is needed is a dialogue between governments, companies, green consumer organizations, and other related actors to work towards a plan of action to deal with managing computers from a life cycle perspective. Researchers and analysts need to continue the process of developing input to inform the process of formulating public response.

5.1 Non-technical Summary

Governments, companies, and civil society are increasingly taking action to manage the end-of-life handling of computers. While appropriate treatment of the waste stream via recycling is required, it is also important to address the flow upstream through reduction of final demand by extending the lifespan of computers. The short lifetime of a computer is not due to it wearing out or breaking down—usually a consumer buys one from the desire to use new software or new functions that require up-to-date hardware. The extension of lifespan is thus not about making computers more durable, but rather the meeting of users demand for computing power with fewer new machines. Reselling computers to secondary users and upgrading performance to match current specifications are two avenues to do this.

This chapter addresses the question of the relative effectiveness of reselling, upgrading, or recycling a computer to reduce environmental burdens. The fact that much of the energy-related and other environmental impacts of a computer occur in the production phase suggests that the extension of lifespan could be particularly effective at reducing burden (i.e., fewer new machines implies less impacts to produce). Also, recycling of computers will not "refund" environmental burdens to the extent it does for many other products. This is because much of the environmental impacts of making of computer are in realizing its complex form, not its bulk substance.

The environmental effectiveness of resell, upgrade, and recycle options for computers is evaluated through quantitative estimation of energy use in these three cases. Results indicate that reselling or upgrading one in ten computers reduces total energy use by 8.6 percent and 5.2 percent, respectively (by reducing demand for new machines). In contrast, recycling the materials in one in ten computers only saves 0.43 percent (by replacing demand for virgin materials). The difference is dramatic and suggests that extension of lifespan should receive real attention on the policy agenda addressing end-of-life computers.

Economic issues, user demand, and their knowledge need also to be considered when considering how policy and other social responses can extend computer lifetime. Computer recycling remains an expensive proposition; recent legislation in Japan mandating collection and recycling of computers will cost the nation's consumers some hundreds of millions of dollars per year. Encouraging reselling and upgrading could be a cost-effective way to enhance the overall system. Vibrant used-PC markets in the United States and Japan indicate substantial consumer demand, but difficulties in transferring software licenses to the secondary users and lack of awareness among general users are two obstacles to its growth (see Chapter 10). A variety of responses from governments and company actions could correct these (and other) impediments to reselling and upgrading.

REFERENCES

Amore, D. 1998. Study finds computer recycling not clicking. *Waste Age* 30(12): 14–15.

Curran, M. 1996. *Environmental life-cycle assessment.* New York: McGraw-Hill.

Das, S. and S. Matthew. 1999. Characterization of material out-puts from an electronics demanufacturing facility. 1999 IEEE International Symposium on Electronics and the Environment, 11–13 May, at Danvers, Massachusetts.

Dillon, P. 1998. Technical report #6: Potential markets for CRTs and plastics from electronics demanufacturing: An initial scoping report. Technical report #6. Chelsea Center for Re-cycling and Economic Development, University of Massachusetts. <http://www.chelseacenter.org/Publications1.asp> (17 May 2003).

European Parliament. 2003a. Directive 2002/95/ED of the European Parliament and of the council of 27 January 2003 on the restriction of the use of certain hazardous substances in electrical and electronic equipment (RoHS). *Official Journal of the European Union,* L 37:19–23.

———. 2003b. Directive 2002/96/EC of the European Parliament and of the Council of 27 January 2003 on waste electrical and electronic equipment (WEEE). *Official Journal of the European Union,* L 37/24.

Green Design Initiative. 2003. Economic input-output life cycle assessment. Pittsburgh: Green Design Initiative, Carnegie Mellon University. <http://www.eiolca.net/> (17 May 2003).

Hendrickson, C. T., A. Horvath, S. Joshi and L. B. Lave. 1998. Economic input-output models for environmental life-cycle assessment. *Environmental Science and Technology* 32(4): 184A–191A.

Japan Electronics Industry Development Association (JEIDA). 2000. *White paper on personal computers 2000–2001* (in Japanese). Tokyo: Japan Electronics Industry Development Association.

Japan Times, The. 2002. Panels agree on new recycling scheme for PCs. *The Japan Times*, 22 February. <http://www.japantimes.com/cgi-bin/getarticle.pl5?nn20020222a9.htm> (17 May 2003).

Lifestyle Center. 2002. *Statistical databook of the IT Society 2002* (in Japanese). Tokyo: Seikatsu Jouhou Center.

Miyamoto, S., M. Tekawa and A. Inaba. 1998. Life cycle assessment of personal computers for the purpose of design for environment (in Japanese). *Energy and Resources* 19(1): 75–80.

Pepi, J. 1998. Technical report #7: Scrap electronics processing. Chelsea Center for Recycling and Economic Development, University of Massachusetts Amherst. <http://www.chelseacenter.org/Publications1.asp> (17 May 2003).

Takesue, H. 2000. Assessment of environmental impact in the recycling of communications equipment. Proceedings of the Fourth International Conference on Ecobalance, 31 October–2 November, at Tsukuba, Japan, 555–558. Society for Non-Traditional Technology.

Williams, E., R. Ayres and M. Heller. 2002. The 1.7 kg micro-chip: Energy and chemical use in the production of semiconductors. *Environmental Science & Technology* 36(24) (15 December): 5504–5510.

Williams, E. and Y. Sasaki. 2003. Energy use in sales and distribution via e-commerce and conventional retail: A case study of the Japanese book sector. *Journal of Industrial Ecology* 6(2): 99–114.

Chapter 10

TODAY'S MARKETS FOR USED PCS—AND WAYS TO ENHANCE THEM

Eric Williams[a] and Ruediger Kuehr[b]

[a] *United Nations University, Japan*
[b] *United Nations University, Zero Emissions Forum, Germany*

1. INTRODUCTION

The environmental impacts associated with the production and disposal of personal computers are exacerbated by their short lifespan, which increases demand for the production of new units and, ultimately, adds to the number of computers destined for landfills or recycling centers. Extending the lifespans of computers should therefore be a priority in their environmental management. One important and practical way to do this is by encouraging markets for used personal computers (PCs). Computers are normally disposed of long before they break down or wear out; rather, the user wants a new machine with better performance and new functions. Not all users require high performance, however; the most popular applications of PCs (e-mail, Internet, office software) often work just as well on older machines. Despite falling prices in recent years, PCs remain an expensive item; thus, presuming it can meet their computing needs, many users will find the lower price of a used machine attractive.

The used-PC market is primarily driven by economic forces; there are, however, "non-economic" obstacles to its reaching its full potential. The environmental and social benefits delivered by the used-PC market imply that efforts should be made to ensure that Adam Smith's "invisible hand" can do its work. The first step in maximizing the flow of used PCs is to understand their markets. This chapter will examine the origin and structure of the demand for used PCs, the status of markets and the supply chain ("re-supply" may be a better word), and identify issues relevant to encouraging future growth.

Computers and the Environment: Understanding and Managing Their Impacts
Edited by Ruediger Kuehr and Eric Williams, pages 197 – 209.

2. DEMAND FOR USED COMPUTERS

Personal computers are multi-purpose devices. The required specifications of a machine depend on its intended use—a key dynamic in the used market. In this context it is useful to clarify the structure of demand—what computer functions are needed or desired by what kinds of users. One simple categorization breaks users into four groups according to two basic characteristics: office versus home use and "power" versus normal use. Power users employ computers for hardware-intensive tasks, such as image/video editing and video games, as well as the usual e-mail, Web browsing, etc. Normal users are mainly interested in only the latter applications, which do not require a state-of-the-art machine to perform well. Although it is a simplified representation of a complex mix of user demands, this picture should be sufficient to extract useful lessons.

In addition to the distinction between hardware requirements, power and normal users often have different levels of knowledge regarding computers. This point is particularly relevant when considering upgrading and reselling options. A less-knowledgeable user may not be aware of the advantages or means of reselling. Also, some may be hesitant to sell or donate their PC because they are not sure how to completely erase their private data from the system. This knowledge-gap is no doubt shrinking as society-at-large becomes more familiar with information technology, but it remains a barrier in the short term.

The distinction between home and office users is important. Most office work is not so demanding of computing power, so from a performance perspective one could expect significant demand from businesses for used computers; but larger firms may be reluctant to opt for used machines unless a trusted vendor can provide a reliable and uniform set of machines. Companies often have information technology (IT) specialists managing their system, lending a standard of expertise in matching computing needs and demand often not present with home users.

A quantitative description of the numbers of different types of users would be helpful in identifying the scale of the potential market for used computers. The relative number of home users, both power and normal, can be estimated for Japan based on surveys of consumer motivation for purchasing a computer. Hardware-intensive applications—games/entertainment and music/graphics hobbies—were reported as the major motivation for 30 and 35 percent of purchasers, respectively (Nikkei 2000). Thus, the fraction of power home users is set at 35 percent, though this is a crude estimate. For business users, a survey taken in April 1999 indicates that 13 percent of computers purchased by businesses had 400 MHz Pentium II processors or better (ibid.). Considering the state of technology at that time, it is reasonable to assume that these computers are intended for power users. These results are combined to yield a quantitative picture of user demand, as depicted in Table 1.

As their computing demands can be well met by older machines, normal users are expected to be the largest potential market for used computers. Given that power users tend to purchase new computers over shorter cycles than normal users, much of the flow of used machines is expected to be from power to normal users. The combined fractions of power and normal users from Table 1 are 21 percent and 79 percent, respectively. As the flow of used computers from power to normal users is expected to be especially important, the resell market may be capped by limited supply (of desirable machines). The substantial fraction of leased computers in the office sector is promising for the used market. Leasing companies are charged with delivering a given service for a fixed time period at a reasonable cost, so they should be interested in reducing costs through utilizing and selling used equipment whenever possible.

Table 1: Shares of different categories of users in Japan.

Category	Share	Sub-category	Share	User type	Share	Overall share
Office	55%	Lease, rent	51%	Power	13%	4%
				Normal	87%	24%
		Purchase	49%	Power	13%	4%
				Normal	87%	23%
Home	45%			Power	35%	13%
				Normal	65%	32%

Source: Adapted from JEIDA (2000) and Nikkei (2000).

3. PRICES OF NEW AND USED COMPUTERS

Needless to say, the attractiveness of a used PC depends crucially on its price relative to a new machine. It is not only the price of the base hardware that is important, but that of the entire system, including software and warranty. The average price for a new desktop PC system in 2002 was U.S.$801 (Spooner 2002). Table 2 shows more recent prices of new and used, low-end desktop systems. System 1 was chosen as a typical option for a normal user purchasing a new computer. New computers for power users cost anywhere from $1,500 to $3,000, depending on the configuration. Systems 2 and 3 illustrate the price for refurbished computers; the latter sacrifices some performance for lower price. System 4 shows that previous-generation technology suitable for a normal user is very inexpensive, but this system does not come with an operating system (OS) or applications. This is often the case for online vendors and brokers, for reasons to be explained later. Local resellers probably offer similar products, with the operating system and applications installed, for around $350. Shipping a computer by courier can easily run into the tens of dollars; thus shipping costs become a significant fraction of lower-priced used systems. This suggests that local resellers enjoy a slight economic advantage, particularly for low-end machines. But on the other hand, local sales limit the potential market and may only perform well in centers of high population. In some cases, the export of used equipment to industrializing countries may be more profitable (due to higher demand for low-priced equipment), despite added shipping costs. For example, the refurbishing firm MIREC, located in a more rural area of Southern Germany, reports that it finds local demand for its products insufficient, and thus has developed a business selling used equipment to Eastern Europe.

Table 2: New and used PC prices.*

	System	Price (U.S.$)
1	Dell Dimension 2350 (new): Celeron 1.7 GHz, 256 MB RAM, 60 MB HD, DVD drive, 17" CRT monitor, XP Home, WordPerfect suite, 3-year warranty	$808
2	Dell Dimension 2350 (refurbished, vendor: Dell): exact specs as above	$703
3	Dell Dimension 2350 (refurbished, vendor: Dell): Celeron 2.0 GHz, 128 MB RAM, XP home, WordPerfect, 1-year warranty	$460
4	Gateway P3 (used, Internet vendor) Pentium III 550 MHz, 128 MB RAM, 17" CRT monitor, No OS or applications, 30-day warranty	$210

* As of February 2003. Shipping not included.

The difference in prices between new and used was more pronounced in the past; various factors have resulted in dramatic drops in new PC prices in the last few years. It is not clear how far this trend will continue in the future, but in general, the trend toward lower prices for new PCs hurts the used market. In general, the more expensive a product is, the more robust the secondary market; consider automobiles and houses, for example. To sum up, under current market conditions a "do-it-yourself" used PC with previous-generation specs is much cheaper than a new machine. Opting for a refurbished machine with software and warranty included significantly reduces the price gap between it and a new computer, but is still relatively inexpensive.

4. STATUS AND GROWTH OF USED-PC MARKETS

One study put the total value of the global market for used IT equipment at $9 billion in 2001 (Lei 2002). This figure is no doubt rough, as national statistics (such as the U.S. Census) do not have a category for used IT equipment, and there is as yet no industry organization or consulting firm that regularly surveys the industry.

For the United States, the International Data Corporation published a study in 1998 of the used-PC market (Luening 1998). According to their estimates, the total market in 1997 was 5.5 million units, with 14 percent annual growth. The study also predicted that the growth in the used market would decline to 10 percent per year due to competition from lower-priced new PCs. Apparently, there has been no follow-up analysis to determine if this actually happened. One estimate put the scale of domestic sales of PCs (including used) in the United States at 30.3 million machines in 1998 (Dyrwal et al. 2000), suggesting that the used market has around an 18 percent market share in unit sales.

Mic Research Institute Inc. estimates the Japanese market in 2001 for used computers at 830,000 machines and on track for 18 percent annual growth in 2002 (My-Com PC Web 2002). By comparison, the Japanese market for new PCs in 2001 was 12 million units, down 11 percent relative to 2000. Presuming both U.S. and Japanese estimates are reasonable, it would seem that the Japanese used market is much smaller than in the United States. There are many possible factors that might explain this; the reputed aversion of Japanese consumers for used goods is sometimes suggested by analysts as the main cause. The prices of used PCs are somewhat higher in Japan (System 4, for instance, runs near $300), but new PCs are somewhat more expensive as well, so there is little relative difference.

No estimates of used-PC markets in regions other than the United States or Japan were found in the literature search of English, Japanese, German, and French language sources. The lack of data on Europe is especially unfortunate, as the European Union is both a major consumer of IT and the most active in implementing regional-level legislation to deal with waste electronics. The only related information found was an Internet survey with some 2,600 German computer users reporting (Stiller 2001), where 54 percent of respondents said they have sold (at least once, presumably) their old computer to the used market using Web ads or auctions, and 32 percent said that they had bought a used PC. This suggests that, in Germany at least, there may be an active used market facilitated by consumer-to-consumer sales via the Internet.

The existing and potential influx of used IT equipment to industrializing countries is also important. While there are many factors contributing to the digital divide, the high price of IT hardware is no doubt important. Domestic and international trade in used computers presents an avenue to stimulate the dissemination of IT in the industrializing world. But imports of end-of-life electronics is a double-edged sword—the positive potential also carries with it the risk of becoming a dumping ground for waste equipment from rich countries. A worst-case scenario of this risk is exemplified by the environmental carnage incurred from the local computer dismantling and materials recovery industry in the Chinese town of Guiyu (Puckett and Smith 2002). How this trade can be handled so as to allow useful imports (i.e., usable, inexpensive used PCs), but prevent dumping is a question that has yet to be addressed.

5. SOURCES OF USED PCS

It is worth mentioning at the outset that the structure of the used-PC supply chain, like all secondary markets, is different in nature from that of new products. There is no producer as such; rather the supply comes from the consumption sector itself. Various business intermediaries offer services (such as information, warranties, and transport) that enable the flow between primary and secondary users.

The main potential sources of PCs for the used market are companies, homes, and the government and education sectors. One would naively expect an efficient flow of PCs from larger companies to the used market, because resources and expertise are presumably available to ensure wise management of capital investments. Firms leasing computers reportedly sell off-lease machines[1] to the secondary market (Fishbein et al. 2000), these computers may represent a major part of the supply to the used market. Anecdotal evidence suggests that companies that bought their computer will sometimes donate it to charity at its end-of-life (after storing for some amount of time), but rarely sell on the used market. Home users have a financial motivation to promptly sell their old machines on the used market. Some surely do, but a number of sources suggest that most home users store their old computer for some years and then throw it away (Goldberg 1998; Matthews et al. 1997). After a few years in a closet, a computer is no longer attractive to most potential secondary users. Based on anecdotal evidence, it appears that governments and educational institutions probably do not sell to the used computer market in general. Proper surveys, however, would do much to help understand the real extent to which various groups supply the used market, but none exist as yet.

6. INTERMEDIARIES IN THE USED-PC MARKET

The different agents involved in facilitating the flow of used PCs to second users can be classified as resellers, brokers, and donation agencies.

6.1 Resellers

Resellers are agents that physically handle the used PC, possibly checking functionality, refurbishing, installing software, and/or offering warranties. A bricks-and-mortar reseller is a traditional shop where customers can examine and try products before purchasing. Customers of these local shops are, reputably, usually small- and medium-size businesses. The largest chain of resell shops in the United States is the Computer Renaissance franchise, which has 110 stores in North America.[2] Not surprisingly, reselling PCs over the Internet is very active; a plethora of small- and medium-size firms are engaged in the business. Original equipment manufacturers (OEMs) are also in the game; Dell, HP/Compaq, and IBM sell their own refurbished machines online to both corporate and private customers. These manufacturers generally do not handle previous-generation technology (Pentium III or lower); they are apparently targeting the market for "high-end" refurbished goods.

6.2 Brokers

A broker, in this context, is an intermediary who simply provides information to potential secondary users on available machines. Classified print ads are the traditional

---

[1] A computer lease is typically between two and three years, tending to be closer to two years (Fishbein et al. 2000)

[2] <www.computerrenaissance.com>

avenue for used markets, and now, since the advent of the Web, there are also Internet classified ads. Local ads have the advantage that they introduce nearby buyers and sellers, reducing transport costs. The Internet auction expands the market of buyers and sellers, and has proved a popular and successful model. eBay, the largest Internet auctioneer in the United States, facilitates a booming trade in IT equipment, valued at about $2 billion in 2001, of which 46 percent is used equipment, 14 percent refurbished, and 40 percent new (Keafe 2002).

6.3 Donation Agencies

A number of organizations around the world have been set up to act as facilitators for donations of used IT equipment. Generally, the market value of the machines, warehousing, and shipping costs are tax-deductible. In the United States, the Cristina Foundation focuses on donation (and training) to disabled, learning, and poor persons.[3] The Computers for Schools network provides matchmaking between donors and local schools.[4] Wayne Tosh of Computers for Schools says they estimate that 400,000 used computers have been placed in schools in Canada.[5] There are also organizations specializing in donations to industrializing nations, such as the African Computer Literacy Project (Ghana) and African Regional Counsel for Computing (Kenya).

The above description identifies the different organizations involved, but does not analyze the key issue of who are the current and future major players. These are important questions for the future of the used-PC industry, but difficult to answer given the lack of information on the sector. The number of computer manufacturers selling refurbished products has increased in recent years, suggesting that they are taking the market more seriously. This could be due in part to the shrinking profit margins on new machines and the increased input of sale-able machines from improved company-level take-back systems.

7. DEMAND FOR USED PCS

Like the supply, the different customers for used PCs can be divided into companies (small, medium, and large), home, and government and education sectors. Analysts identified education and small business sectors as the main purchasers of used PCs in the United States in the late 1990s (Computer Dealer News 1998). Cost considerations are no doubt a major factor, though for small business the flexibility of a smaller organization may also be an important factor. While consideration of computing needs suggests that medium and large companies could benefit from used computers, new PCs are probably more easily managed in the context of a larger

[3] <www.cristina.org>

[4] <www.pcsforschools.org>

[5] <www.schoolnet.ca/cfs-ope>

organization. For instance, system management of a large network is no doubt simpler for uniform groups of new machines, rather than a diverse mix of used PCs. Schools are obviously a significant market, with a large number of users to be provided with basic computing functions and (in general) very limited budgets. Governments are apparently not significant customers for used PCs at present, but this situation could change if the purchase of used PCs was made a higher priority in purchasing guidelines. Home users are a significant market, but this market is likely still below its potential size. The high-end refurbished market is probably accessible to the home user; consumers can buy these machines online from OEMs. Efforts to purchase second-generation technology in today's market can quickly become confusing, however. Which generation of machine suits my needs? Which resellers can I trust? Which ones will offer machines with the operating system and desired applications already installed at a reasonable price? What is a realistic price? How does one find the going market value for a machine? These are difficult questions for many potential home buyers, who may thus opt for a new PC just because the choice is simpler, though more expensive.

8. OBSTACLES TO GROWTH OF THE USED-PC MARKET

8.1 Software Licensing

As mentioned in Section 3 on prices of new and used computers, many resellers (and donation agencies) offer computers without any operating system or applications installed or included. This significantly reduces the attractiveness of the machine to many potential buyers. Preparing it for use involves extra cost and trouble, especially compared to new PCs, which are usually shipped ready with operating system, Web, e-mail, and office applications installed.

Given that software is an integral part of the product, why do many resellers offer computers with no software? The primary cause is related to the licensing agreements of software that is packaged with new computers. Most licenses for pre-installed software allow a user to transfer the right to use along with the ownership of the original hardware. This right is rarely exercised in practice, however, due to a number of factors. To begin with, hardly anyone is aware of the nature of end-use licenses to begin with. To properly transfer the right, the original owner must also pass on the printed copy of the agreement that comes with the computer. Many users do not save this document, partly because it is usually a small piece of paper separate from other documents that are obviously useful, such as the user manual. Even if the former user does pass the license agreement to the next user, it is often difficult to tie the document to a particular computer, putting the reseller at legal risk (at least it is so perceived by them). Microsoft is known to be strict on licensing issues and reportedly patrols eBay to check larger resellers advertising there for possible license infringements (Kannellos 2001).

8.2 User Knowledge and Supply/Demand Issues

Many consumers are not aware of existence of the used market, either as a place to sell their old computers or as a potential place to purchase another one. They may be also concerned about data security and unaware of the ways to ensure that their data is properly removed from the hard drive. Besides, they may not know that older-generation technology is entirely suitable for their computing needs.

8.3 Inadequate Standardization of Used-PC Prices

To ensure smooth trade, buyers and sellers should know the appropriate price of a product. Either side may balk at a deal if there is significant uncertainty on price. "Blue books" (price guides) can be an important tool to reduce this uncertainty. Blue books play an especially important role in used markets, as it is often difficult for consumers to shop around for prices on a comparable product. For computers this problem is exacerbated by the vast array of configurations available; the value of a given model varies considerably depending on its installed memory, storage devices, and software. There are blue book pricing services for computers. For example, www.orionbluebook.com hosts the standard Web-based tool for such as service. The cost of a quote for a single model is $4, a price many potential buyers and sellers are probably unwilling to pay. In addition, this quote does not account for the variety of possible configurations of a computer.

8.4 Taxation System

The taxation system also has an effect on the used market. Under the German taxation system, for instance, firms can count 100-percent of the costs of a computer and its peripherals as depreciation within a three-year period after purchase. Even private PC users can count depreciation of 35 to 50 percent of the purchase price if they have proof that a certain amount of use is for work. For regions with high taxes, users are economically rewarded for replacing their PC at least every three years. Although the service delivered by a used PC may be the same as a new one, the taxable deductions are far larger for the latter.

9. PROMOTING USED-PC MARKETS

9.1 Easing the Transfer of Software Licenses

There are two directions to pursue to improve the inclusion of software with used PCs on the market. One is to ensure that the license for pre-installed software does indeed get transferred to the next user. Raising awareness of the issue for users, producers, and resellers would lead to more licenses being saved and transferred. The licenses themselves can be clearly tied to an individual computer by printing its serial number on the document, and made easier to save by attaching it to the main user manual (e.g., in a pocket on the cover) or even to the computer itself. Producers would be encouraged to do this if the practice were one of the requirements for a computer to obtain an eco-label. Current computer eco-labels, such as the Blue Angel and the E.U. Flower, do not address software issues at all.[6]

9.2 The Open-Source Option

Another direction to pursue is the use of "open-source" software.[7] Linux, the most well-known open-source operating system, has evolved into a user-friendly package with a graphic-user interface. OpenOffice is an open-source package of office applications—including word processing, spreadsheets, and e-mail—available for Windows, Mac, and Linux operating systems.[8] It is easy to install and use, delivers most of the functions available in commercial applications, and can handle documents made with Microsoft Office. It is thus possible to equip a computer partly or entirely with open-source software, which allows it to deliver the main functions demanded by normal users. Some compatibility problems with Microsoft Office documents remain, however, and these systems are not yet well known or trusted enough to gain wide acceptance.

9.3 Raising User Awareness

While there has been some coverage, the popular media has yet to really bring home the message to consumers that extending the lifetime of their computer is something they can do that benefits both the environment and their pocketbooks. It should be in the interest of the media to cover the topic, as people are generally interested in both helping the environment and saving money. Publicly-funded support

---

[6] See Chapter 8 for more on computer eco-labels.

[7] "Open-source refers to software whose license requires that its software code be open, extensible, and freely distributable. It also allows for collaborative development, which refers to the worldwide community of software developers participating in a continuous cycle of development, review, and testing over the Internet." Open-Source Software Institute <http://www.oss-institute.org/>.

[8] <www.openoffice.org>

for awareness raising campaigns could be considered if interest from commercial media is insufficient.

9.4 Improved Blue Book for Used PCs

An inexpensive, flexible, and well-known pricing service would help to stimulate the used-PC market. Such services exist for the used automobile market, but not yet for PCs. It may be that the high price of existing used-PC blue book services arises from relatively low demand; the price of a knowledge service generally goes down with increased demand. This would imply a "chicken-and-egg" dilemma; without a de-facto pricing standard many potential buyers and sellers are reluctant to enter the market, yet without the market an inexpensive pricing intelligence cannot be supported. If this is the case, temporary publicly-funded support could be given to establish a blue book service.

9.5 A Used PC-Friendly Tax System

It is possible, without a doubt, to rearrange tax-deduction systems to reduce the financial incentive for consumers to replace PCs with new machines. A simple solution is to use a fixed deduction per computer bought, regardless of the actual price. Another option is to give a more generous valuation of used PCs or a longer depreciation period (five instead of three years). The former would increase incentive to donate used PCs to charity while the latter would reduce the tax incentive for firms to frequently purchase new machines.

9.6 Increased Use of Leasing (Possibly)

Increased leasing of computers could help to ensure a timely flow of computers to the used market, but the environmental benefits of this are not yet clear. The key point is that while leasing firms are quite active in selling their off-lease machines, this does not automatically insure an increase in total useable lifespan of the machine. If the "lease lifespan" is significantly shorter than the average life of a purchased office computer, the second life of the resold off-lease machine may not be long enough to realize a net gain. The average lease term is two to three years (Fishbein et al. 2000), while the average lifespan of purchased office computers is reportedly 3.4 years (Smulders 2001). Leases can be extended beyond the initial term, however, and the leased computer is sometimes purchased by the lessee. The net change in lifespan is unknown and there is insufficient information at present to resolve the question. If it turns out that leasing does increase the net lifetime of computers, increased practice of leasing by firms and governments can be recommended as a policy to reduce the environmental burdens of computers.

9.7 Visible User Fees in End-of-life Legislation

There are many national and regional initiatives in place and under development to legislate the collection and recycling of end-of-life IT equipment. Of these, the WEEE Directive of the European Union is perhaps best known.[9] At the very least, such collection systems should be designed so as not to negatively affect used markets, and should include incentives for users to take advantage of the resale market. One possibility is to charge a visible fee to the consumer for computer recycling. This would stimulate consumers to avoid the fee by selling their computers on the used market. Such a fee was planned for the Japanese take-back system for launch in the summer of 2003, but was later switched to a non-visible fee charged by manufacturers (Aritake 2002).

9.8 Gathering Intelligence on Secondary Markets

Used products tend to slip through the nets of national statistics. This is partly because much of the trade is informal (consumer-to-consumer), but even the formal sales by companies are missed, because census reporting routinely lumps new and used products together. Even given a lack of official statistics, economic activities of any significant size usually have an associated organization or group of organizations devoted to gathering data and developing intelligence to aid the development of the sector. This is decidedly lacking for secondary IT hardware, probably since the main agents are small firms, donation agencies, and individuals, who do not have the luxury of paying for expensive reports from consulting firms. Governments have long been known to support analysis of certain industries; for instance, yearly surveys by the U.S. Geological Service have made them a major source of intelligence for mineral industries. Thus far, no agency, at least in the United States, the European Union, or Japan, has been charged with keeping track of the used-IT market. Given that so many governments are regulating the end-of-life of IT equipment, it also makes sense that they should invest in ensuring that fewer computers end up in landfills or recycling centers in the first place.

This analysis is one of the very few addressing the used-PC market—it may be the first publicly-available study to proactively consider how the sector itself is to be promoted. Future work based on better market data and surveys of important market players can surely do a better job than this one of identifying the opportunities. Given these opportunities, it is hoped that society will make efforts to take advantage of them in order to fully utilize IT equipment and reduce its environmental burdens.

[9] Directive 2002/96/EC of the European Parliament and of the Council of 27 January 2003 on waste electrical and electronic equipment (WEEE).

ACKNOWLEDGEMENTS

The authors thank William Cade, Robin Ingenthron, Yvette Marrin, and Wayne Tosh for providing helpful information for this chapter.

REFERENCES

Aritake, T. 2002. Japan to require recycling of PCs. *International Environment Reporter* 25(6) (13 March): 267.

Computer Dealer News. 1998. Used PC market continues to evolve. *IDC Review and Forecast* 14(16) (27 April): 20.

Dyrwal, Janine, Jose Rodriguez and Jean-Yves Moreno-Guzman. 2000. *ReVAMP IT: A study on the international used computer resale opportunities*. Washington, D.C.: Kogod School of Business at American University. <http://www.american.edu/carmel/jd1000a/index.html> (18 May 2003).

Fishbein, B., L. McGarry and P. Dillion. 2000. Leasing: A step towards producer responsibility. New York: INFORM, Inc. <http://www.informinc.org/leasingepr.php> (18 May 2003).

Goldberg, Carey. 1998. Where do computers go when they die? *The New York Times*, 12 March. <http://gdi.ce.cmu.edu/comprec/nytimes98/12die.html> (18 May 2003).

Japan Electronics Industry Development Association (JEIDA). 2000. *White paper on personal computers 2000–2001* (in Japanese). Tokyo: Japan Electronics Industry Development Association.

Kannellos, M. 2001. Microsoft cracks down on used Windows sales. *CNET News.com*, 9 October. <http://news.zdnet.co.uk/story/0,,t269-s2096850,00.html> (18 May 2003).

Keafe, Collin. 2002. Where haves meet wants. *Dealerscope.com* (January issue): 51–54. <www.dealerscope.com>.

Lei, T. 2002. Refurbished IT goods take off. *Asia Computer Weekly*, 8 April.

Luening, E. 1998. Used PC market surging. *CNET News.com*, 31 March <http://news.com.com/2100-1001-209684.html> (18 May 2003).

Matthews, H. Scott, Francis McMichael, Chris Hendrickson and Deanna Hart. 1997. Disposition and end of life options for personal computers. Carnegie Mellon University Green Design Initiative technical report #97–10. Pittsburgh, PA: Carnegie Mellon University. <http://www.ce.cmu.edu/GreenDesign/comprec/NEWREPORT.PDF> (18 May 2003).

My-Com PC Web. 2002. 2001 market for used computers at 830,000 units (in Japanese). *My-Com PC Web*, 22 August. <http://pcweb.mycom.co.jp/news/2002/08/22/23.html> (18 May 2003).

Nikkei. 2000. Nikkei market access annual IT basic data 600, (in Japanese). Tokyo: Nikkei Publishing.

Puckett, J. and Ted Smith, eds. 2002. Exporting harm: The high-tech trashing of Asia. Seattle, WA: Basel Action Network and San Jose, CA: Silicon Valley Toxics Coalition, 25 February. <http://www.svtc.org/cleancc/pubs/technotrash.htm> (18 May 2003).

Smulders, C. 2001. Watching Rome burn: PC empires threatened by extending life cycles. Doc. AV-14-1972. Stamford, CT: Gartner Inc.

Spooner, J. 2002. PC sales see a ray of light. *CNET News.com*, 2 August. <http://news.com.com/2100-1001-948115.html> (18 May 2003).

Stiller, A. 2001. Schnäppchen aus zweiter Hand, Viel PC für wenig Geld. *C't* (November):104.

ACKNOWLEDGEMENTS

REFERENCES

Chapter 11

RECYCLING PERSONAL COMPUTERS

Stefan Klatt
MAN Nutzfahrzeuge AG, Germany

1. INTRODUCTION

Electrical and electronic equipment, such as televisions, video recorders, hi-fi systems, and refrigerators, are a part of almost every household today—especially in the developed world. And in shops and offices, electronic cash registers, bar code scanners, computers, fax machines, and copiers have become indispensable. As a consequence, the amount of waste electronic equipment flowing from households and businesses has increased dramatically.

Information technology equipment is a significant part of the growing volumes of electronics waste. Every second year in Hanover (Germany) at CeBit, the world's largest computer fair, the heart of the problem is evident in the colourful on-screen presentations, new mini-computers, colour printers, and electronic gadgets—the booths are full of the latest models, while models that were brand-new only two years earlier are already outdated. As one can see with personal computers (PCs), technological and product developments are occurring at a breakneck speed.

People wanting to dispose of an "outdated" PC (which may still be functioning) might store it at home, throw it out as waste or, in the rare exception, take it to the municipal recycling station. The bulkier the old computer system, the more likely it is that the consumer will try to trade it in with a dealer when buying a new one—but then the dealer has to do something with it.

The growth in PC production has added to the increasing volume of industrial waste, further reducing the remaining capacity of landfills to critically-low levels, as documented in Japan, for example (Soeda 1999). The disposal of PCs along with household or commercial waste is no longer acceptable. Waste from electronic products contains a wide array of toxic substances (albeit in relatively small amounts), some of which can easily escape into the air or groundwater if disposal is not handled properly. According to the United States Environmental Protection Agency (EPA),

Computers and the Environment: Understanding and Managing Their Impacts
Edited by Ruediger Kuehr and Eric Williams, pages 211–229.

electronic equipment waste is the single largest contributor of heavy metals to the U.S.'s waste stream, aside from lead-acid automotive batteries. An estimated one-third of the lead found in municipal solid waste in the year 2000 was attributed to electronics waste (Arensman 2000).

The following are among the potentially hazardous materials found in electronics waste (including old TVs, PCs, printers, and household appliances) (Arensman 2000):

- Lead - used in solder and for radiation shielding in cathode-ray tubes (CRTs)
- Cadmium - used in batteries
- Antimony - used as a flame-retardant, chip encapsulate, and a melting agent in CRT glass
- Beryllium - used in connectors in cell phones and in older PCs
- Chromium - used in metal plating operations
- Mercury - used in the bulbs that illuminate flat-screen displays (in very small amounts)

In Japan in 1997, for example, the volume of discarded PCs totalled 41,000 tonnes (metric tons), 30,000 tonnes from businesses, and 11,000 tonnes from households. In response, the Japan Electronic Industry Development Association recently announced that it is committed to establishing a PC recycling system, ranging from recovery and recycling to sharing associated costs (marketing, overhead, etc.). The association pointed out the need to promptly establish a recycling system in view of the anticipated increase in used PCs discarded by households. They are planning to expedite the establishment of a recycling system for household PCs in cooperation with autonomous bodies and recycled-PC dealers, as well as by keeping close links with the recycled-PC market (*Japan Chemical Week* 1999).

The U.S. EPA cites European studies that estimate that the volume of electronics waste—including old TVs, PCs, printers, and other aging high-tech scrap—is growing 3 to 5 percent a year, almost three times faster than the overall municipal waste stream in Europe, which is estimated to be growing at 1 to 2 percent a year. Very little electronics waste, particularly from consumers, is being recycled. Industry's recycling effort is still aimed almost exclusively at products collected from its commercial customers, who tend to replace equipment frequently, when it still has relatively high resale or salvage value.

Today, key parties, including manufacturers and recycling companies, mainly discuss organizational questions—which technologies to use and (most importantly) who will pay for them. In the meantime, a policy framework—already discussed for years—is still unrealised in most countries. But new hope for regulations that will ensure equal competition determined by standard quality requirements is offered by initiatives such as the European directive on the recycling of scrap electronics (WEEE), which formulates basic approaches for the relationship between product development and environmentally-oriented recycling. The responsibility and costs for the disposal of products from private households are to be handed over to manufacturers and local authorities. It is expected that this shift in responsibility from the public domain to the manufacturers will lead to improved recycling of old equipment and, through direct feedback, will lead to the development of innovative recycling-oriented products.

2. QUANTITIES AND MARKET STRUCTURE

Computers and other electronic equipment contribute only about 1.6 percent to the total annual production (1997) of hazardous waste (214 million tonnes) in the United States (Yu 2000). Electronic industries are placing greater importance, however, on national economic competitiveness, and have a growing recognition that many processes employed to produce electronic products have environmental consequences. These considerations have led the industry to proactively examine environmental practices. Some electronic products are facing increasing disposal problems. For example, despite improved recycling efforts, there will still be 55 million PCs landfilled by 2005 in the United States, according to a report from Carnegie Mellon University (Matthews et al. 1997).

Compared to the 36.7 million PCs shipped from manufacturers in the United States in 1998, the percentage of PCs recycled was only about 6 percent. In contrast, for major appliances, including washing machines, water heaters, air conditioners, refrigerators, dryers, dishwashers, ranges, and freezers, the ratio of units recycled to units shipped was about 70 percent in 1998 (Amore 1999).

The constantly increasing figures have not yet reached a peak. The growth rate for PC sales to the private sector is enormous. As a consequence, the potential for future scrap electronics is equally large.

Additionally, the average lifespan of information technology (IT) equipment is decreasing drastically due to the enormous speed of developments in the field. While a lifespan of ten years was common in the past, IT equipment is being recycled today after only five to six years. The lifespan of equipment bought today will be only 2.5 to 4 years. According to these facts, a tripling of the amount of scrap IT equipment will have to be dealt with in the near future (bvse 1998).

In spite of the positive prognosis for the recycling industry with the growing quantity of used-IT appliances, the amount of businesses engaged in recycling scrap electronics has decreased noticeably in recent years, as in Germany, for example. There are many reasons for this development.

- The coming into force of new environmental legislation has not yet been able to prevent further landfilling and incineration of old appliances. Furthermore, there is a lively export of scrap electronics; experts talk of 60 to 80 percent of Germany's scrap electronics being exported, for example (EUWID 1999). A recent study of The Basel Action Network (BAN) and the Silicon Valley Toxics Coalition (SVTC) describes the trade with e-waste as an export of real harm to the poor communities of Asia (Puckett and Smith 2002).
- For years, product-related legislative initiatives have been intensively discussed without leading to any kind of concrete result. Confident that something would happen, the industry has made considerable investment in disposal and recycling capacities. These plants are working far below their capacity.
- In the meantime, some multinational disposal groups have tried to edge out some smaller recycling companies in the market with dumping offers. Operators

of big plants try to work to capacity by all means. The already existing ruinous pricing policy (not reaching the break-even) is thus further supported.

- Many manufacturers have resorted to building on their own recycling capabilities. Modern information and communication technologies promote extensive globalisation and integration of products and markets, benefiting large companies mostly (bvse 1997).

In spite of this development, some medium-size businesses have managed to perform successfully in the national German market. Often they are attractive to a select group of customers because of their services in logistics and on-site dismantling. Due to their knowledge of materials and their processing experience, they can also create lucrative business opportunities in the reselling of recycled materials. The trend towards a joint approach and disposal service is also evident. Even established companies are trying to break away from mere electronics waste recycling and enter new fields in servicing.

3. LOGISTICS

In the process of collecting and recycling electrical and electronic appliances (including old TVs, PCs, printers, and household appliances), logistics generally make up more than 50 percent of the total processing cost (INTERSOH AG 1997). Factors driving up these costs are

- the multitude of potential sources of waste generation;
- the multitude of appliances, differentiated by size, weight, construction, and harmful substances they potentially contain;
- costly options for adequate collection strategies, like curbside collection and collection point systems; and
- the expenses of establishing and operating transfer sites where goods collected are sorted and redirected for processing.

Dividing electronic appliances into large-, medium-, and small-sized categories has proven to be useful in terms of logistics (Waste Magazine 1999). Today, small appliances, to a large extent, are not separated and recycled, but are landfilled or incinerated as household waste.

Sorting waste into secondary resources and residue, however, is an absolute requirement for the recovery of waste, according to the recycling economy law (Waste Letter 2000), which has resulted in the demand for the separate recovery and environmentally-oriented recycling of electronics waste in all current proposals for legal regulations.

Most experts believe that the problem for the electronics industry has not yet "hit the street," saying that we have become an "attic" society, not a "throwaway" society. They refer to the millions of old PCs stored in people's basements, attics, and garages. By one estimate, 125 million PCs will end up in U.S. landfills by 2010 (Bartholomew 2001).

In terms of user-friendliness, there is a distinction between curbside collection and collection point systems. Curbside collection has the advantage of its popularity, resulting in high collection rates, but it demands higher expenditures for logistics. The collection point system—in which used equipment is carried by the user to central facilities, where it is sorted and redirected for further handling—represents a relatively high expenditure for the generators of the waste. As a consequence, the consumers (that is, the ones who choose not to simply store their used equipment) tend to dispose of their used electronics through the collection system for household waste, even if the equipment is relatively large (e.g., a monitor).

It can be established that curbside collection systems guarantee the best return rates due to their high degree of user-friendliness, but extensive expenditures are required for logistics. It can be integrated into the process. Thus, synergies could be utilised and additional investments could be minimised.

4. RECYCLING PROCESSES

Other than products in consumer electronics, IT appliances contain a high percentage of metals; that is why, in comparison, they are more attractive and thus more valuable for recycling. In addition, there is an increasing demand for used components from IT equipment. A basic requirement for recycling is the removal of toxic substances, particularly from batteries and accumulators.

Recycling is more economically beneficial if it gives priority to product recycling (i.e., re-use of complete appliances, components, and modules) rather than recycling of the materials that make them up. However, because only certain appliances are suitable for complete product recycling, the recycling of the materials in them is of major importance. In the process, the different materials that are contained in scrap electronics are recovered and re-used as secondary resources in the production of new products.

4.1 Input

The percent of equipment received in a typical recycling pilot project is shown in Table 1 (Jung 1998).

In the numbers of computer equipment collected, central processing units (CPUs) accounted for 35 percent and monitors accounted for 33 percent. Laptops only represented 1 percent of the total mix of equipment collected. Other pilot projects indicated that, by weight, CPUs and monitors comprised over 82 percent of the electronics waste stream (Chelsea Center 1998a).

Table 1: Computer equipment received in pilot project.

Type of equipment collected	Percent (%)
CPUs	35
Monitors (black & white)	33
Printers	15
Keyboards	12
Peripherals (mouse, hard drives, cables, etc.)	2
Miscellaneous parts (circuit boards, fans, etc.)	2
Laptops	1
Total	100

Source: Jung (1998).

4.2 Demanufacturing and Technical Reprocessing

Demanufacturing can be visualised as a multi-step process (Das and Matthew 1999). At each step, one more part with certain commonalties is separated or removed. The separated element could be a partial part, an entire part, or a subassembly of parts. The separated subassembly may be further disassembled into parts and materials.

Typically, as we progress down the disassembly line we observe three phases. In the first phase, the disassembler attempts to improve the accessibility to different parts of the product. As a result, no value is directly released in this phase. In the second phase, valuable parts and subassemblies are reclaimed. Finally, in the third phase, separation is done to facilitate downstream material recycling processes. This ensures a higher grade of input purity to the material shredder, and consequently results in a better quality recycled material.

The PCs are then moved on to mechanical processing (i.e., to be disassembled by hammer and cutting mills). In the next step, metal parts are separated by means of conventional separation techniques (vibration, cyclone, sink-float, turbulence, magnetic separator, air flow, screening machines, etc.).

The ferrous metals collected are reprocessed in traditional steel-waste processing and smelting plants. Non-ferrous metals, to the extent that they can be isolated, can be used and processed as secondary raw materials. By using cutting mills, parts are mechanically turned into a mixed granulate, from which up to 99 percent pure copper can be reclaimed through physical separation methods.

Obviously, this is also valid for the processing of printed circuit boards that contain plastic and metal compounds as well as precious metals. After pre-grinding, high-quality circuit boards are processed in a specialised non-ferrous metal works, and the individual metals are separated by means of electrolysis. Today, even poor-quality circuit boards can be mechanically separated from their aluminium and iron parts in special plants. The heavy metal compounds, together with the carrier material (resin paper, fibreglass), receive the same treatment as the high-quality circuit boards with a higher amount of precious metals. Due to the lower treatment costs, compared to the costs and wages in the iron and steel industry, it is often tried to separate this mixture of materials mechanically. Yet, this procedure usually causes residual wastes (fine fragments) that are difficult to dispose of.

To sum up, the problems associated with reprocessing metals and metal compounds contained in electronics equipment (including their printed circuit boards) can generally be addressed by existing methods. Reprocessing is also lucrative economically.

4.3 Disposal

In case safe reprocessing in accordance with environmental regulations is not possible or cannot be carried out due to capacity reasons, the materials must be disposed of in accordance with the public interest. Doing so, the choice is between the disposal of the material through thermal treatment or disposal in landfills. Still, considering environmental standards, this is the worst solution.

Components like LCD displays, condensers, and other non-separable compounds containing contaminants are sent through the usual disposal processes. In case they contain mercury, it can be recovered using a reprocessing method. The recycling of small parts that are not in the most common categories of plastics, metal, or compounds (e.g., batteries, fax drums, and printer drums) is not done in most cases, because of high costs and the possibility of using ecologically questionable alternatives. Figure 1 shows a typical demanufacturing and disposal process.

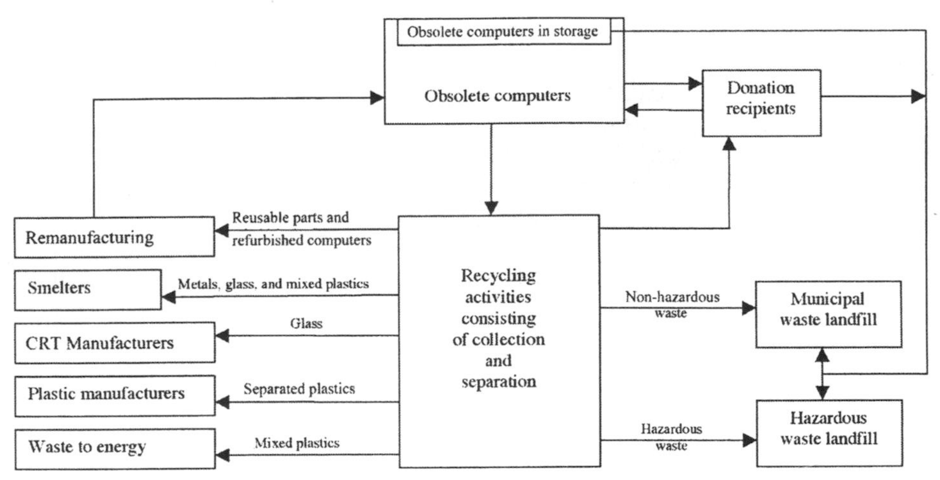

Figure 1: Flow chart of recycling processes.
Source: USGS (2001).

4.4 Output

Ideally, the goal of material recycling is to economically generate material that is no different from virgin material. But in the manufacturing and assembly process the virgin material is inevitably combined with a variety of different materials, and in a sense, becomes impure. A product composed of a single material (e.g., beverage can or glass bottle) provides the highest recycling yield, since the impurity levels are very low. A level of high impurity in recycled materials is one of the most common obstacles to more widespread recycling. The level of purity determines the quality of the recycled material and its market price (Das and Matthew 1999).

The outgoing electronics composition[1] typically consists of the materials shown in Figure 2 (Chelsea Center 1998a).

One of the objectives of the disassembly process is to separate the materials contained in used equipment and then place them in different bins for each type of material, so that each bin has a high level of homogeneity or purity. Clearly, the disassembly effort and costs increase as effort is made to improve the level of purity.

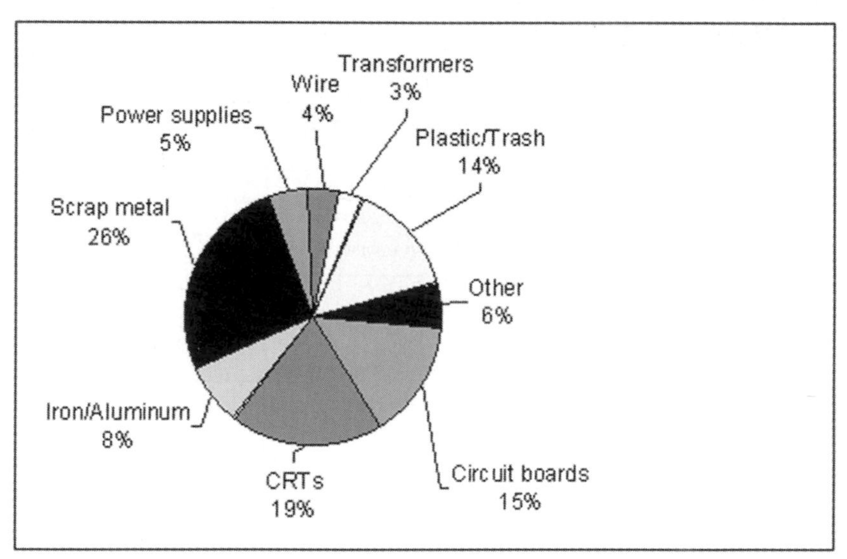

Figure 2: Outgoing electronics composition.
Source: bvse (2000).

The design and operation of an efficient electronics demanufacturing facility (EDF) requires well-defined and, potentially, standardized output bins. These bins must be defined in terms of their primary content, impurity levels, and generation rates. This information is also of importance for design for environment (DfE) analysis. Product

[1] Average proportion of composition of materials in new IT products.

design must attempt to minimise the impurity levels and make the recovery of larger masses of homogenous material easier. The most important output components (steel, aluminium, copper, circuit boards, precious metals, batteries, cathode ray tubes, plastics, glass, etc.) are outlined in the following section.

- Steel is commonly found in medium- and large-sized electronic devices. Typically, the product chassis and support plates are fabricated from steel. For example, 30 percent by weight of a mainframe computer or large server is steel. Steel is relatively easy to reclaim because of its unique magnetic properties. A steel recycler will shred the bin contents and then use a magnetic separator to remove the steel.
- Aluminum is an expensive material that enjoys strong demand in the recycling industry. The percentage of aluminum in computing equipment and consumer electronics is generally relatively low, but components such as disk drives do contain significant amounts of aluminum. An EDF will typically maintain two aluminum output bins—one for high and one for low quality material. For high quality, the impurity level is restricted to 2 percent. High quality aluminum is refined to produce industrial grade material. For aluminum mix,[2] the allowed impurity is in the 10 to 20 percent range. The material is shredded and then passed through an eddy current separator, which improves the purity, although the process is slow and expensive. The reclaimed aluminum material is then melted in a rotary furnace. Because of the added processing costs required to produce useable material, the price of aluminum mix is less than half of high-quality aluminum.
- Copper is a common material in most electronic products, but most of it is difficult to completely isolate. An EDF will typically maintain two copper output bins, one for high and one for low quality, as in the case of aluminum. For high quality copper, the impurity level is limited to 1 percent. The recycled material vendor wants this commodity to be close to pure copper in quality, since it is melted into grade "A" copper. High-quality copper is a valuable output from an EDF, since buyers are willing to pay a premium price. For copper mix, the allowed impurity level is in the 20-to-30-percent range, and varies greatly between vendors.
- Printed circuit boards are frequent outputs of the de-manufacturing process. Most come from personal computers, mainframe computers, copiers, and modems, etc. Approximately 25 percent of circuit boards have a high content of precious metals. This requires the EDF to have a detailed knowledge of the product's contents, in order to identify the valuable chips. Apart from halogenated flame retardants, mostly integrated into the base material, other components that might contain hazardous materials include batteries and accumulators, as well as condensers, sometimes even mercury relays and LCD displays.
- Electronic boards with a high amount of precious metals are sent to precious metal refiners. Here, all types of precious metal are recaptured for the

[2] Aluminium mixed up with other materials (copper, plastics, etc.).

manufacturing of new products. Circuit boards and batteries contain most of the heavy metals, and circuit boards contain the highest precious metal values. Information on the specific amounts of individual precious metals (gold, silver, and the platinum group metals) recovered is unavailable, but as much as one-third of the precious metals recovered from scrap may be gold (Bleiwas and Kelly 2001). The value of the one tonne of precious metals recovered from electronic scrap in the United States in 1998 probably exceeded $3.6 million. This value was obtained by assuming that the precious metals recovered consisted of 60 percent silver and 40 percent gold, palladium, and platinum. One tonne of circuit boards can contain between 80 and 1,500 grams of gold and between 160 and 210 kilograms of copper (Veldhuizen and Sippel 1994, as cited in USGS 2001). This is 40 to 800 times the concentration of gold and 30 to 40 times the concentration of copper contained in ore mined in the United States. Gold in an obsolete computer has little or no value. Like ore, it must be collected, concentrated, and refined in order to acquire a high value.

Unfortunately, as computer manufacturers strive for lower costs, they are reducing the amount of precious metal in the manufacturing process (reducing the useful life and dependability of the electronics as a by-product). This reduces the value of the recycled electronics—making the entire recycling process less economical.

- Batteries and accumulators of many designs are used in electrical and electronic appliances to allow operation without being connected to an electrical power supply and to ensure that data is secure in the event of a power failure. By law, all batteries must be removed during the recycling process (Yu 2000). This is a time-consuming process, since the product may have several batteries installed at different locations inside the device.
- Computer monitors and televisions contain cathode ray tubes (CRTs), which, because of their large content of lead and barium and small amounts of cadmium—dangerous heavy metals—constitute a serious environmental problem in landfills.

Plastic housings are removed from the CPUs and printers in order to access and recover the valuable materials and components inside. The plastic housings currently go to a local landfill for disposal. Markets for mixed plastic resins are limited, but several market options exist or have potential, including

- granulated mixed resins for use in asphalt paving products
- sorted and granulated, sold to product manufacturers
- manufacture of low-end, large-piece or thick-walled products such as pallets or lumber with mixed resins
- Glass represents about 9 percent of electronics scrap. Ordinary glass in electronic products is easy to recycle and free of major technological or environmental problems.

4.5 Technical Barriers and Obstacles

Common problems still faced in the recycling of electronic products include the lack of data about the material content of products that enter the recycling process and methods of recycling lead, printed circuit boards, plastics, and glass from cathode ray tubes (Chelsea Center 1998b). Obstacles and technical barriers, especially concerning plastics and CRTs, will be discussed next.

4.5.1 Lack of product data

A primary difference between the processes of manufacturing and demanufacturing is the lack of product data. Most EDFs, even in-house operations, have limited knowledge about the design and material composition of products they handle. Given that a primary objective of the EDF is to disassemble and sort the parts into material bins, this is major handicap.

4.5.2 Metal Recycling

Metals comprise 57 percent of the total amount of electronics scrap. Metals used are primarily iron, cast iron, stainless steel and other steel alloys, aluminum and aluminum alloys, copper alloys, lead, and zinc. Among these, lead is commonly used in manufacturing CRTs and soldering. There is no cost-effective alternative for lead yet. Research is ongoing in several organizations to develop lead-free solder and glass alternatives.

4.5.3 Printed Circuit Boards

There are a large number of environmental issues involved in printed circuit board manufacturing and recycling. The manufacture of printed circuit boards alone brings not only waste but also many environmental problems, such as hazardous materials disposal, pollution generation, and energy and water consumption.

4.5.4 Plastics Recycling

Plastic materials represent 19 percent of total electronics scrap. One of the problems with plastics recycling and recovery is the difficulty in getting high-quality material. While thermoplastic materials are recyclable, compound plastics are unacceptable. Because of the incompatibility of various plastics, the parts disassembled from old products must be identified and sorted into separate different types of plastic. Further studies are necessary to establish (1) the most appropriate treatment for each type of plastic (recycle, regenerate, or incinerate), (2) effective and economically viable identification techniques, (3) more effective separation technologies, and (4) approaches for efficient co-mingling of virgin and recycled plastics.

There are processing costs associated with the currently available markets for mixed plastic. Processing costs could decrease, perhaps to no cost, if local manufacturing applications for mixed resins could be further developed and marketed.

Co-mingled plastics are more difficult to find markets for than single resin materials and have a lower market value. One final option, therefore, is to identify and sort the plastics, either manually or using plastic separation and cleaning technologies. The biggest issue is not whether it is technically feasible to sort, but whether it can be done economically.

Several companies in the United States, including Recycling Separation Technologies (RST) in Massachusetts, have perfected processes for separating mixed plastic granulate into single resin products. These companies can take whole plastics housings, grind and granulate the plastics, mechanically separate resin types (e.g., density separation), and remove metal, rubber, and paper contaminants to produce a clean feedstock. These single resins can be introduced into a greater variety of applications than mixed resins, even replacing virgin materials in high performance electronics applications.

So far, plastics from used electronic appliances are often reprocessed energetically (i.e., incinerated and used for heat generation). Usually, this is because only small amounts of specific plastics are obtained at each individual dismantling plant, which cannot be treated for materials recycling due to economic reasons. In view of the environmental aspects, however, materials recycling would have advantages. A possible solution is currently being developed, based on a feasibility study conducted on behalf of the Ministry of Environment and Transport Baden-Wuerttemberg in Germany (Ministry of Environment 1998). In the course of this study, from among the various used appliances, a certain kind of plastic was identified which can be obtained from many recycling plants. As a result, it may be recommended for use in future appliances, on condition that it should be rather easily separable from the used appliances and add up to an amount significant enough to be of interest to manufacturers, given the appropriate logistical conditions.

4.5.5 Recycling CRT Glass

In televisions, the picture tube (cathode ray tube) accounts for 55 percent of the weight; in PC monitors, it is approximately 32 percent. They contain toxic fluorescent materials and consequently have to be treated as hazardous waste. That is why it is necessary to implement appropriate methods for a high-standard treatment of the picture tube glass. A number of different mechanical, thermal, and chemical recycling processes for picture tubes are available on the market.

Colour picture tubes consist of two different types of glass. The screen is made with glass containing barium or strontium, whereas the reverse side of the picture tube contains lead. Furthermore, unlike monochrome tubes, colour tubes have a metal shadow mask. During the recycling of this kind of tube, the fluorescent coating has to be removed, and additionally, the different sorts of glass and metal parts have to be separated. Many obstacles need to be overcome before recycling glass from PC monitors and TV panels can be considered as a serious option, including the separation of lead-containing glass from the recycled glass and obtaining reliable analyses of the contents of the glass at a reasonable cost.

Markets currently exist for both used equipment (i.e., whole monitors and televisions) as well as CRT glass (Chelsea Center 1998b). For whole CRTs, some equipment is more desirable than others. Some markets accept some types of cullet (broken glass from used appliances) but not others. The lead content of CRT glass is the most important factor determining its potential and appropriateness for use in any given application. The lead content and chemical characteristics of CRT glass vary by component as well as by manufacturer.

The following are the current markets for CRTs and CRT glass:

- Refurbishment and resale
- CRT intermediate processors
- CRT glass manufacturing
- Industrial lead panels
- Decorative glass manufacturing
- Export
- Smelters

There are two possible options for using recycled panel glass: (1) use the recycled glass as a raw material, much like sand or feldspar, or (2) use the recycled glass as cullet, which must be of the same quality as domestic cullet (Hermans et al. 2001).

Using the recycled glass as a raw material means that certain conditions must be met:

- All lead-containing glass and other potentially harmful contaminants must be effectively removed.
- The material must be ground to a grain size of less than one to two millimetres to be able to mix it with the other batch ingredients and enable reliable analysis of the material.
- The material must be representatively sampled and analysed, and maintain a stable colour and the same working point and density as the glass that is to be produced.

Option two—using the recycled material much like domestic cullet—seems to be an attractive way to recycle panel glass. To use this option, however, recyclers need to be absolutely certain about the chemical composition of the cullet material, since the coarse grain size makes it difficult to obtain a representative sample.

Using the "process as cullet" option means

- properly identifying each individual panel/funnel combination with respect to glass composition and colour components;
- separating panel and funnel, and grinding away the lead-containing parts of the seal edge;
- carefully cleaning the panel of harmful contamination (phosphors, pins, and masks);

- breaking the panel glass into the desired grain size, primarily between <5 and <25 millimeters, depending on the batch house equipment; and
- sieving off the excess fine amount (typically less than 10 to 20 percent by weight of the lot—fine dust will cause foaming if present in large amounts).

The individual identification of each panel presents a problem. Unlike glass from food containers, TV panel glass compositions have shown quite a large variation in the past decades. There is no reliable, fail-proof, and internationally-accepted way to identify types of glass. Analysis of each individual panel (e.g., by X-ray fluorescence) is technically possible but cost-prohibitive. What is needed is a cheap and reliable analysis technique, able to process large quantities of samples with little pre-processing.

Finally, it should be noted that not all the TV glass that comes from the market can be processed for TV glass production. The maximum recycling rate is limited and will remain so in the coming decades. Alternative uses are needed for the glass that comes from recycling, such as for building and road construction materials.

4.6 Costs

The economics of computer recycling actually depends on a number of factors (Jung 1998), including

- the location of the equipment in relation to the location of the equipment processors and end-users;
- current market conditions for computer equipment parts and materials;
- the age of the equipment;
- the volume of equipment;
- the material composition of the equipment (e.g., precious materials content, use of materials compatible for disassembly, sorting and/or melting);
- the technology used by the equipment processor (e.g., dismantling versus shredding, infrared sorting versus manual sorting); and
- markets used by the equipment processors (e.g., local, international, simple or sophisticated).

Additionally, there are significant differences regarding the recycling costs in industrialized and non-industrialized countries. For example, CRTs are not considered hazardous in many undeveloped countries and a higher demand exists for reusing whole computer monitors, so fewer handling, processing, transportation, and disposal costs would be incurred.

In a research project at the University of Massachusetts, quantitative data was collected on the processing rates, labour times, and market values of the materials from the university's electronics demanufacturing program (Chelsea Center 1998a). By weight, CPUs and monitors comprised over 82 percent of the institutional electronics stream (Table 2). The average direct labour cost for electronics dismantling was $119 per tonne (19.1 hours per tonne at $6.25 per hour), and the variation around that average cost for various component types was minimal. It is estimated that the measured dismantling time represents about 85 percent of the total labour involved in the operation. Total labour time would include time spent loading and unloading containers. If this additional labour were included in the total direct labour cost, the total cost would be approximately $140 per tonne.

The revenue for the study period (net of CRT recycling and trash disposal costs) was $175 per tonne of material processed (Table 3). It was estimated that contract administration, utility, and supplies costs were under $35 per tonne. If the $140 per tonne in direct labour costs is combined with the above miscellaneous costs it appears that program revenues offset its expenses. As a result, the university realised a benefit of $70 per tonne in avoided disposal and hauling costs.

Table 2: Incoming electronics processed by weight, time, and cost (U.S.$).

Unit Groupings	Weight (kg)	Time (minutes)	kg per minute	Hours per metric ton	Dollars per metric ton
Audio visual	129	162	0.80	20.9	$131
CPUs	5,749	7,206	0.80	20.9	$128
Monitors	5,455	6,934	0.79	21.2	$132
Keyboards	32	33	0.96	17.3	$108
Miscellaneous	230	299	0.77	21.6	$135
Parts	148	185	0.80	20.9	$131
Printers	1,835	2,259	0.81	20.5	$129
Totals	13,578	17,078	0.80	21.0	$128

Source: UMASS (1998).

Table 3: Outgoing commodities: composition and value (U.S.$).

Commodity name	Kilograms of product	% of outgoing commodities	Rev./Expense cents per kg	Dollar revenue or expense
A boards	670	6.0%	2.2	$1,475
Aluminium	225	2.0%	0.44	$99
Backplanes	104	0.9%	3.3	$344
CRTs	2,095	18.8%	0.22	-$461
D boards	969	8.7%	0.121	$117
Disk drives	195	1.8%	0.308	$60
Fans	42	0.4%	0.132	$6
Goldfingers	1	0.01%	35.2	$32
Goldchips	1	0.01%	46.2	$42
Iron/aluminium	910	8.2%	0.198	$180
Scrap metal	2,882	25.9%	0.0385	$111
Power supplies	564	5.1%	0.088	$50
Wire	450	3.6%	0.33	$149
Transformers	283	2.5%	0.154	$44
Yokes	149	1.3%	0.11	$16
Plastic/trash	1,592	14.3%	-0.077	-$123
Totals	11,131	100%	5.57	$2,141
			Dollars / ton	$175

Source: UMASS (1998).

During the San Jose Computer Collection and Recycling Pilot it was found that collecting and recycling used consumer computer equipment through retail stores is feasible, but can be costly (Jung 1998). The fact, however, that dramatically different costs for recycling computer monitors were found in the study demonstrates that there is not a singular set of economics for computer recycling. Due to the significant difference in monitor recycling costs, two net economic scenarios were calculated for the San Jose pilot. The net economics for the first scenario (with the monitors recycled in China) came to a program cost of $4,373, or $142 per ton. The net economics associated with the second scenario (monitors recycled in the U.S.) reached a cost of $17,990, or $584 per ton.

The cost for recycling the computer monitors ranged from $0.05 to $0.50 per pound, while the net economics for collecting and recycling all collected computer equipment ranged from a total cost of approximately $4,400 to $18,000 ($0.07 to $0.29 per pound). Although the overall cost for recycling was substantial, it was still lower than the costs that could have been encountered if the computer monitors had been disposed of as hazardous waste and the remaining equipment had been landfilled.

In summary, no one has yet figured out how to turn a profit on the entire process of electronics recycling, especially when collection and transport costs are included.

5. RE-MANUFACTURING USED COMPUTER TECHNOLOGY

The outcome of all investigations into the material flows in the waste industry can be simplified to one statement: The more expensive a product is when produced and the shorter its service lifespan, the more damage it does to the environment. Still, by remanufacturing certain parts and components, the ecological damage as well as the production costs can be reduced. If this strategy is pursued single-mindedly, the apparent contradiction between "additional costs through measures for environmental protection" and "the pressure to reduce costs in order to maintain competitiveness" can be solved. The short cycles of innovation here are both a challenge and an opportunity.

The first objective of an EDF is to either remanufacture the product for a second life or to reclaim and re-use valuable parts or subsystems. Unfortunately, only a small percentage of disposed products can presently be realistically remanufactured or have parts reclaimed. The majority must enter either a material recycling stream or be landfilled. Product redesign and technological obsolescence precludes many remanufacturing efforts. For instance, the nickel-cadmium and lithium-ion batteries in cellular phones and laptops could potentially be re-used, but they typically do not fit into new designs.

In mechanical and vehicle engineering, the use of treated and tested second-hand parts and components has a long tradition in the spare parts business. In the IT field this is still the exception, though a market is developing slowly. But the use of second-hand components in new products is a lot more common than one would expect; several companies worldwide are engaged in selling used components in a big way. Purchasers are mostly from cost-sensitive sectors in the Far East (e.g., entertainment electronics), like manufacturers of musical greeting cards or toys, for instance, which do not make great demands on the quality of components.

A central problem with the remanufacturing of electronic parts and components is the guarantee required of the quality and reliability of the components.

The area of remanufacturing is gaining importance, and not just from a manufacturer's point of view. Due to a stagnating market, businesses that recycle scrap electronics are trying to find new opportunities to reduce costs while producing additional profits. Electronic data processing (EDP) systems and PCs in particular constitute an excellent potential for remanufacturing. These EDP systems are found, for instance, in administrative departments of industrial businesses, as well as in banks and insurance companies. Here and there, appliances that have been in use for only two years are taken out of service, because the use of modern networks (with usually demanding software) often makes it necessary that the existing system is not only adapted and expanded, but has to be replaced as a whole.

6. CONCLUSION

In the autumn of 2000, the Sixth Session of the Conference of the Parties to the Framework Convention on Climate Change took place in The Hague (the Netherlands). One of the main messages at the meeting was that it is time to take action, and thus, the topic of sustainable development is now on everyone's lips—one would hope. Computers are becoming as ubiquitous around the world as televisions. But as newer, faster, and cheaper PCs are churned out, where do the obsolete systems go? Many studies prove that relatively few old PCs are being recycled. Instead, most are stored in warehouses, basements, or closets, or have met their end in municipal landfills or incinerators, although there is currently increased attention on the environmental impact of recycling scenarios used for end-of-life electric and electronic products. An issue that has surfaced in recent years is the need to shift from a world of mass production to one where keeping materials and resources in circulation supports both economic growth and environmental protection. Recycling is the key to achieving this goal.

Meanwhile, all participants continue to pay increased attention to waste product recycling systems. In many of the efforts to date, however, the focus has been mainly on reducing the volume of waste landfilled and on the recovery of valuable materials, and less on several technical barriers and obstacles, such as the lack of product data and the recycling of printed circuit boards, plastics, picture tubes, and cathode ray tube glass. A few aspects of these technical barriers, such as recycling problems, environmental impacts, and potential markets, have been discussed thoroughly.

The use of second-hand, treated and tested parts and components has a long tradition in a few businesses. In the IT field this is still the exception due to current methods of product redesign and technology obsolescence, but the area of remanufacturing is gaining importance. Electronic data processing systems and PCs in particular constitute an excellent potential area for remanufacturing.

This chapter shows that the economics of recycling PCs depend on a number of factors, such as local logistics, current market conditions, used recycling technology, and the material composition of the equipment. Additionally, we have found significant differences regarding the recycling costs in industrialized and undeveloped countries. Some studies prove that there are dramatically different costs for recycling computer equipment in different regions and circumstances. There is no singular set of economics for computer recycling.

In conclusion, it is essential that discussion continues on the environmentally-sustainable treatment of scrap electronics. There is no shortage of technical possibilities for the recycling of scrap electronics, which can, at the same time, be ecologically justifiable and economically lucrative. Many powerful, usually medium-sized, businesses could take advantage of these opportunities, if they were provided the necessary basic conditions to support growth in this field.

REFERENCES

Amore, Dawn. 1999. Study finds computer recycling not clicking. *Waste Age* 30(12): 14–15.

Arensman, R. 2000. Ready for recycling? *Electronic Business* (the management magazine for the electronics industry), November.

Bartholomew, D. 2001. Where have all the PCs gone? *Industryweek.com*, 5 March.

Bleiwas, D. and T. Kelly. 2001. Obsolete computers, "gold mine," or high-tech trash? Resource recovery from recycling. U.S. Geological Survey Fact Sheet 060-01.

Bundesverband Sekundärrohstoffe und Entsorgung e.V. (bvse). 1997. *Entwicklung des Elektronikschrottrecyclings aus der Sicht mittelständischer Unternehmen* (Progress in electronics scrap recycling from the point of view of medium sized businesses). November. Bonn: bvse.

———. 1998. *Elektronikschrottrecycling - Fakten, Daten, Zahlen* (Electronics scrap recycling—Facts, figures and procedures). Bonn: bvse.

Chelsea Center for Recycling and Economic Development. 1998a. Scrap electronics processing, technical report. Chelsea, MA: University of Massachusetts Amherst.

———. 1998b. Potential markets for CRTs and plastics from electronics demanufacturing, technical report. Chelsea, MA: University of Massachusetts Amherst.

Das, S. K. and S. Matthew. 1999. *Characterization of material outputs from an electronics demanufacturing facility*. Newark, NJ: Dept. Of Industrial & Manufacturing Engineering, New Jersey Institute of Technology.

European Economic Service (EUWID). 1999. *Marktbericht Elektronikschrottrecycling* (Market research electronics scrap recycling). May.

Hermans, J. M., J. G. Peelen and R. Bei. 2001. Recycling of TV glass: Profit or doom. Philips display components—glass development. *American Ceramic Society Bulletin* (T19) 80(3).

INTERSOH AG. 1997. *Pilot for collection of electronics scrap*, final report. Cologne. April.

Japan Chemical Week. 1999. Home PC recycling system under consideration. *Japan Chemical Week*, 22 April.

Jung, L. B. 1998. *The San Jose collection and recycling pilot*. The U.S. Environmental Protection Agency, 10 July.

———. 1998. *The San Jose computer collection and recycling pilot*. The U.S. Environmental Protection Agency, Common Sense Initiative, 16 July.

Klatt, S. 2000. Bundesverband Sekundärstoffe und Entsorgung e. V. (bvse). Seminar: Zukunft des Elektronikschrottrecyclings. Bonn: bvse.

Matthews, H. S., Francis C. McMichael, Chris T. Hendrickson and Deanna Hart. 1997. Disposition and end-of-life options for personal computers. Carnegie Mellon University Green Design Initiative technical report. Pittsburgh, PA: Carnegie Mellon University.

Ministry of Environment and Transport Baden-Wuerttemberg. 1998. Aspects for municipals regarding recycling of electronics scrap, August.

Puckett, J. and T. Smith, eds. 2002. Exporting harm: The high-tech trashing of Asia. Report released by the Basel Action Network and Silicon Valley Toxics Coalition, 25 February.

Waste Letter. 2000. Waste and sustainable development. May.

Waste Magazine. 1999. Collection systems.

Soeda, Sh et al. 1999. *Life cycle assessment of a personal computer in consideration of recycling system.* Fujitsu Limited, Japan.

University of Massachusetts (UMASS). 1998. Technical report #7, Scrap electronics processing. Office of Waste Management, University of Massachusetts.

U.S. Geological Survey (USGS) July 2001. Fact sheet. Obsolete computers, "gold mine," or high-tech trash? Resource recovery from recycling. USGS.

Yu, Yue et al. 2000. A decision-making model for materials management of end-of-life electronic products. *Journal of Manufacturing Systems.*

Chapter 12

OPERATIONS OF A COMPUTER EQUIPMENT RESOURCE RECOVERY FACILITY

Joseph Sarkis
Clark University, Graduate School of Management, USA

1. INTRODUCTION

Industrial ecosystems are an integral component of ecocentric and environmentally sustainable management (Shrivastava 1995).[1] For industrial ecosystems to exist sustainably, a "green" supply chain is a prerequisite—including both forward and reverse logistics as primary elements. For products and materials to flow through an industrial ecosystem, the purchasing and logistical functions of an organization are critical inter-enterprise linkages. Purchasing green materials is not necessary for a forward logistics chain, but is necessary for reverse logistics operations to exist. Reverse logistics include a number of internal and external elements, including disassembly, de-manufacturing, and re-manufacturing organizations.[2] These organizations and their customers have both typical and special operations and purchasing requirements. But within a demanufacturing or disassembly organization, the purchasing, operations, and logistics functions have special requirements and characteristics that make it unique, especially from a natural environment perspective.

The goal of this chapter is to provide insight into how a computer demanufacturing facility actually works. It starts with a case study of a resource recovery facility in Contoocook, New Hampshire (U.S.), formerly operated by the Compaq Computer Corporation; it was the first Compaq facility to acquire the environmental management

[1] Industrial ecosystems can be defined as the relationships and flow of products, materials, components, and other resources among various elements of an industrial supply chain, including extractors, producers, decomposers, and consumers.

[2] *Demanufacturing* and *remanufacturing* are related and require that re-usable portions of a product be re-used in the making of a new or partially rebuilt product. De-manufacturing is the process of taking the product apart and determining various useful materials that can be re-used. Remanufacturing focuses more on a "core" part that can be used as central to another product.

Computers and the Environment: Understanding and Managing Their Impacts
Edited by Ruediger Kuehr and Eric Williams, pages 231–251.

standard, ISO 14001 certification. Even though this facility is not typical within the computer industry, there are clear factors that can be generalized to the personal computer (PC) industry and general industry practices, especially those that have significant existing demanufacturing operations or potential.

To set the stage, the chapter begins with a general discussion of green supply chain management and the operational life cycle of a typical manufactured product. The next section provides some company background, including general competitive environmental directions and strategies. The body of the chapter incorporates three major sections. The first will look at the issues surrounding the "inputs" that the Compaq recovery facility faces, including the requirement of purchasing materials from suppliers who, at the same time, are also customers. Some of the general inner workings and findings of this relationship will also be detailed. The second section includes some of the operational characteristics of the facility, including flows and operational process steps, especially those that focus on source reduction and environmental management. Included in this discussion will be the role of the facility's workers and the work environment itself. Since many of the workers are relatively unskilled and on contract (temporary), appropriate training aimed at limiting risk and improving process efficiency will be briefly described. The final section will discuss the outputs of the recovery facility, including some of the issues related to the development of partnerships with potential customers and relationships with suppliers of waste management and recovery services.

1.1 Green Supply Chains and Operations

Competitive pressures have recently started to force organizations to incorporate some dimensions of natural ecosystems into their strategic and operational plans. These new dimensions have not only been evident in traditional industries such as chemicals and petroleum, but have affected the planning and practices of other manufacturing and service-based industries. From a core operational perspective, the natural environment influences the whole value chain—from inbound logistics (where purchasing and procurement play a large role) to manufacturing and operations, to outbound logistics and distribution. The pressure to green has also profoundly influenced support functions such as accounting, engineering, finance, and information systems.

Purchasing and logistical functions are central to the flow of materials through an organization. Purchasing also includes the processes of selecting vendors, designing products and processes, packaging, outsourcing, and inventory management. The relationships among these various elements of the value chain are shown in Figure 1, illustrating the various elements of the green supply chain within a general operational life cycle framework (Bloemhuf-Ruwaard et al. 1995; Lamming and Hampson 1996; Narasimhan and Carter 1998; Sarkis 1995b; Walton et al. 1998). The development of this green supply chain model is based on the typical value chain of an organization (Porter 1985). The primary focus in this figure is the management of products and materials flowing through the supply chain and the relationships among the various functions. Looking at this chain we see that vendors supply the necessary materials that contain varying levels of recycled content. These materials are then stored and may be managed under the auspices of the purchasing function. The production stage incorporates issues such as closed-loop manufacturing, total quality environmental management (TQEM), demanufacturing, and source reduction. Outbound logistics include activities and issues such as transportation determination, packaging, location analysis, and warehousing, as well as inventory management. The external activity—the "use"—is the actual consumption of the product, a situation where product stewardship plays a large role. At this stage, field servicing may occur, but from an environmental perspective the product or materials may be disposed of or returned to the supply chain through the reverse logistics channel. Within this channel, the product can be deemed to be re-usable, recyclable, or remanufacturable. The reverse logistics function may feed directly back to an organization's internal supply chain or to a vendor—starting the cycle again (Carter and Ellram 1998; Fleischman et al. 1997; Pohlen and Farris 1992; Sarkis 1995a). It is this reverse logistics channel that we focus on. The practices of demanufacturing may share similar stages of the value chain as manufacturing (e.g., purchasing, production, logistics), but the products and processes, as well as the practices, appear in a different light.

1.2 Compaq's Commitment

Michael D. Capellas, Compaq's president and chief executive officer put his name to the following commitment published on the company's Web site: "Compaq is committed to conducting its business in a manner that delivers leading environmental, health and safety (EHS) performance and protects the quality of the communities where we operate. We will meet or exceed all applicable EHS regulatory requirements, as well as our own EHS management standards. We will aggressively pursue pollution prevention and waste reduction; encourage re-use and recycling; conserve natural resources; proactively reduce injuries and illnesses; promote healthy lifestyles; and incorporate state-of-the-art EHS practices into our operations and throughout the life cycle of our products and services."[3]

As part of this environmental strategy and mission, Compaq has sought to take care of end-of-life PCs through re-use, recycling, and reduction at their resource recovery facility. Before we begin discussion on the detailed operations at Compaq's resource recovery facility, we first step back and take an overall look at the industry and Compaq's general competitive situation.

[3] <www.compaq.com> (September 2001).

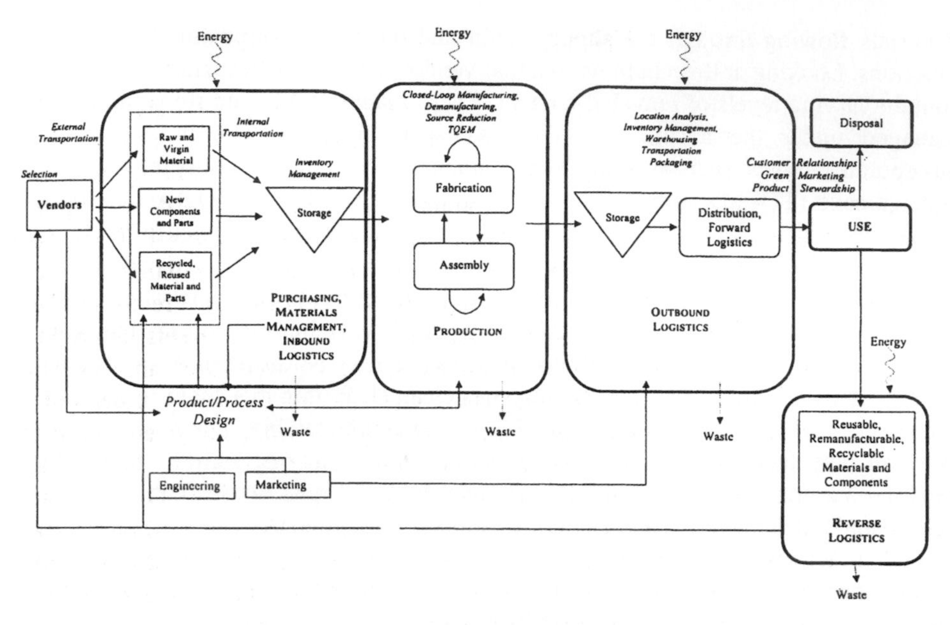

Figure 1: Operational functions and environmental practices within the green supply chain. *Source*: Sarkis (2001).

1.3 The Computer Industry

The computer industry is intensely competitive, with many U.S. and international companies vying for market share. Companies include small clone shops, major components manufacturers, mail-order (Web) distribution, and major multinational corporations. The industry is characterized by frequent technological advances in hardware and software that continue to shrink product life cycles and accelerate rapid introduction of new products, which in turn force the early obsolescence of products and components. Much of the computer industry is driven by integrated systems development. Strategic elements of competition within this industry include distribution (local and global) capabilities, product performance, product quality and reliability, service and support, and pricing. Compaq strives to balance and compete on these strategic elements. To manage the competitive risks, many computer industry companies have also sought to implement technological and marketing alliances with a number of companies, ranging from software giants such as Microsoft, Oracle, and SAP, to Internet service providers like America Online and component manufacturers like Intel.

2. COMPAQ'S BUSINESS

Compaq is a leading global provider of enterprise technology and solutions. The company designs, develops, manufactures, and markets hardware and software solutions and services, including industry-leading enterprise computing solutions, fault-tolerant business-critical solutions, communication products, and desktop and portable personal computers. As of December 2000, Compaq employed 70,100 full-time regular employees and approximately 24,500 temporary and contract workers. Compaq products and services are sold in more than 200 countries directly and also through a network of authorized Compaq marketing partners. It markets its products and services primarily to customers in the business, home, government, and education sectors, operating in more than 200 countries worldwide. During 2000, it derived 55 percent of its revenue from sales outside the United States with its business organized by various geographic regions, such as North America, Asia-Pacific, Japan, Latin America, Greater China, Europe, the Middle East, and Africa. Of Compaq's international revenue, 65 percent comes from Japan, Asia-Pacific, and Latin America.

In June 1998 Compaq completed the acquisition of Digital Equipment Corporation (Digital) for an aggregate purchase price of U.S.$9.1 billion. Digital was an industry leader in implementing and supporting networked business solutions in multi-vendor environments, based on high-performance platforms with an established global service and support team. The reason this acquisition is important for this chapter is that previous to this time, Compaq had outsourced many of its service functions to Digital. Two of the services offered by Digital were it assets and resource recovery capabilities and management of end-of-life products. Compaq, in acquiring Digital, now had a major resource recovery operation in its organizational processes portfolio.

There are three locations within Compaq that carry out "computer asset recovery" of materials from its own products and those manufactured by other organizations within the computer electronics industry. One facility, located in the Netherlands, is the European Materials Recovery Operation. The second is located in Australia (assisted by a New South Wales government waste reduction grant). The third facility, which this case study will examine, is called the Americas Materials Recovery Operation (AMRO), located in Contoocook, New Hampshire in the United States.

2.1 Computer Asset Recovery Within Compaq's Business

Historically, in the 1970s and early 1980s, the AMRO facility (then still owned by Digital) primarily conducted distribution and warehousing functions. In 1986, the product disassembly center was added and eventually became AMRO. Its sole purpose was to serve as an end-of-life facility for Digital products and components. The origination of this facility and its operations were to provide Digital with an avenue to maintain control over its waste stream with a focus on asset management. Initially formed under a reactive policy responding to impending environmental regulations and increased hazardous waste disposal costs, the AMRO operations soon evolved into a strategic element of Digital services, and once acquired, Compaq offered it to its own customer base. Part of the motivation for maintaining its own disassembly facility is that many of the products offered by Compaq are leased products that would return to Compaq at the end of their lease, thus a collection facility was required. Eventually, Compaq realized that many of the products returned as obsolete or irreparable still had components and elements that could be re-used, remanufactured, or recycled. These characteristics allowed Compaq to benefit from the reintroduction or resale of components and materials. What was once viewed as an asset management cost center could potentially evolve a cash flow stream, changing the focus to being a profit center.

Even though this is a relatively nascent business environment, a number of large competitors already exist. Some of the larger competitors are manufacturers of electronic components, such as IBM, Lucent Technologies, Hewlett Packard, and Sun Microsystems. Companies specialized in electronics components materials recovery, such as ECS Refining Inc. and BDI exist, as well. And many "mom and pop" reclamation sites may have specific areas of emphasis in terms of component recovery (some may focus on cathode ray tube recycling, for example).

In this type of industry, the recycling, re-use, and remanufacture of products can still be costly. Joseph Collentro, a former Digital environmental employee, made an observation still pertinent today: For companies to realize the true costs and value of recycling, they need "...to realize that recycling doesn't come for free. The ideal situation will be when manufacturers start including disposal costs with the sales price of their products. At Digital, it cost us around $15 per unit to dispose of a used monitor in a hazardous waste landfill compared to $1 to $1.50 for disposal in a solid waste landfill. Recycling costs currently are comparable to hazardous waste disposal, mostly because of low volume. If more companies use recyclers, costs will probably be more in the $7 to $10-range per monitor, which will include documented tracking for each component as well as recycling. Companies like these offer a service that people are going to demand in the future" (Riggle 1993).

Organizationally, this service now sits within Compaq's hardware services, where it is defined as computer asset resource recovery. Central to this service are the centers which conduct the demanufacturing and warehousing of obsolete and returned computers. They include the following elements in their services:

- Material recovery
- Remarketing—Compaq resells equipment for the customer
- Remanufacturing—Equipment is refurbished for return to the customer or for remarketing
- Parts recovery—Products are disassembled for resale in the secondary market
- Recycling—Compaq handles product disassembly and environmentally safe materials disposition
- Product disposal—For equipment with no market value
- Custom disposal—For meeting special disposal requirements
- Product analysis—To determine time to disassemble, material recovery costs, and material separation characteristics
- Custom reporting—As per local or regional environmental requirements

- Returns management for companies that sell or lease computer equipment (transportation, scheduling, receiving, inspection, and sorting)

Compaq brochures (1999a, 1999b) outline a number of benefits to customers of this service:

- Convenience of partnering with a single provider who provides a full range of integrated options and capabilities
- An easy way to fulfill environmental responsibilities and promote "good corporate citizenship" by customers
- A way to "make the most of" their surplus technology assets with a wide choice of material recovery alternatives
- Flexible revenue-sharing programs that provide for apportionment of net proceeds from recovery
- Assurance of safe, environmentally friendly product disposal
- Solutions tailored to their customer-specific requirements

The AMRO facility, with about 250 employees (approximately 40 percent are contract workers), has been demanufacturing between 25 million and 30 million pounds of discarded electronic equipment every year.

Table 1: Example costs from research report at University of Massachusetts scrap electronic processing center.

Commodity name	Kg of product	% of outgoing commodities	Rev./Expense cents per kg	Dollar revenue or expense
A boards	670	6.02%	2.2	U.S.$1,475
Aluminum	225	2.02%	0.44	$99
Backplanes	104	0.94%	3.3	$344
CRTs	2,095	18.82%	0.22	-$461
D boards	969	8.70%	0.121	$117
Disk drives	195	1.76%	0.308	$60
Fans	42	0.38%	0.132	$6
Goldfingers	1	0.01%	35.2	$32
Goldchips	1	0.01%	46.2	$42
Iron/aluminum	910	8.17%	0.198	$180
Scrap metal	2,882	25.89%	0.0385	$111
Power supplies	564	5.06%	0.088	$50
Wire	450	4.04%	0.33	$149
Transformers	283	2.54%	0.154	$44
Yokes	149	1.34%	0.11	$16
Plastic/trash	1,592	14.30%	-0.077	-$123
Totals	11,131	100%	5.57	U.S.$2,141

Source: Pepi (1998).

The AMRO facility is currently self-sufficient and even generating positive revenue for Compaq. Even though detailed financial numbers are not available for this chapter, we can make some estimates using secondary (exemplary) data from Table 1 of economic data from a small research organization that demanufactures items. For example, using this data, it can be estimated that revenue generated from a 25 million- to 30 million-pound demanufacturing processing plant can be over $2 million annually. It is expected that the profitability is even greater for these larger facilities due to economies of scale. Turton (2000) showed that a demanufacturing plant could generate a revenue stream of $12.6 million with operating costs of up to $10.4 million. To investigate the level of refurbishing necessary to make a demanufacturing facility profitable, Turton provided a generic spreadsheet package to make estimations.

In addition to being profitable, more environmentally significant is the fact that the Contoocook facility is also one of the first 20 organizations in the United States to gain ISO 14001 certification and the first one in Compaq's organization (Table 2).

2.2 Computer Asset Resource Recovery Operations

The AMRO facility serves as a warehouse for returned and excess equipment and as a materials recovery facility. Figure 2 provides a summary of the operational flow of materials through the processes and the personnel or groups executing these processes.

Table 2: Compaq sites that have received or are in the process of seeking ISO 14001 certification.

Site	Year certified	Notes
Contoocook, New Hampshire, USA	1997	Americas Material Recovery Operation Services
Nashua, New Hampshire, USA	1998	Americas Software Manufacturing
Galway, Ireland	1998	EMEA Software Manufacturing
Switzerland Country Subsidiary	1998	Located in Dübendorf, Basel Bern, Geneva, Lausanne, and Rümlang
Ayr, Scotland	1998	Manufacturing
Irvine, Scotland	1998	Service
Erskine, Scotland	1998	Manufacturing
Japan	1999	Manufacturing and distribution
Distribution Center Europe, Netherlands	1999	Manufacturing and distribution
Fremont, California, USA	1999	Manufacturing
Site	**Certification expected**	**Notes**
Singapore	2001	Manufacturing
Riverside, Scotland	2001	Manufacturing

Source: Compaq Web site (2001) <www.compaq.com>.

The operations in the materials recovery portion of the physical plant involve the acquisition and collection of products and materials. At least two types of in-bound flows of material exist. The first flow of material goes directly to the disassembly portion of the facility (typically waste and used products). The second flow of material goes to a sorting section to determine the category of material, which includes separation and inspection of the material by quality control and receiving personnel. The material may be (1) stored as-is in the warehouse facility or (2) tested. The testing section functions like a "triage unit," which determines whether a product is working (re-used, as is), needs some simple "remanufacturing," or needs to be disassembled for recovery of re-usable components or recyclable material. Depending on the material inflow, some materials will require little processing, while others will require a significant amount.

Figure 2: General operations model for demanufacturing facility.

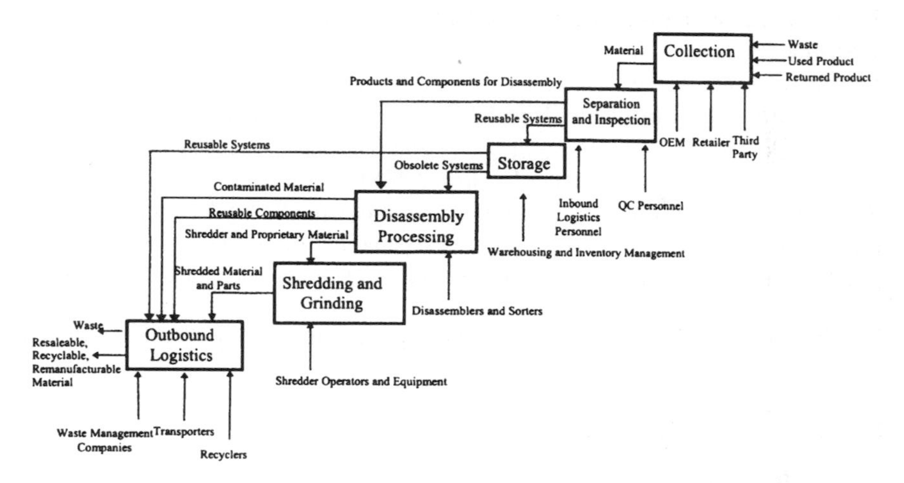

Figure 3 summarizes the sources of the various inputs and types of outputs of this facility. Computer and electronics components form the majority of materials processed within AMRO, but office furniture and other office-based products and materials also arrive for processing. Intact parts are removed, tested, and stocked as spares. Precious metals, (gold, platinum, and silver) are extracted and resold. Plastic, glass, and other materials are recycled. Less than one-tenth of 1 percent of the material ends up in landfills.

Figure 3: Operational and systemic elements of the Compaq Resource Recovery Facility (AMRO).

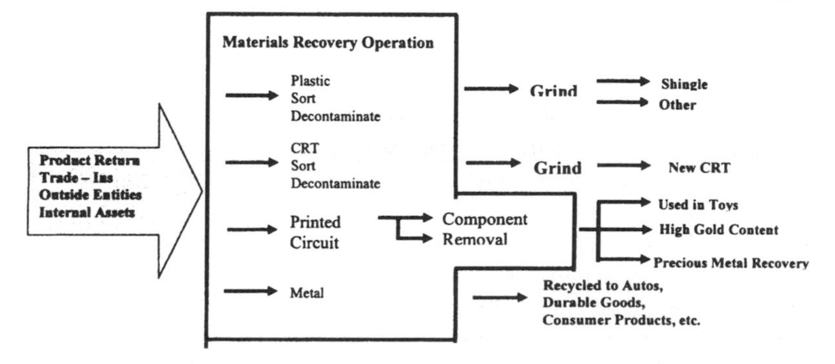

After separation and inspection, some systems are stored because they are re-usable, and then shipped out. Other materials are sent on to disassembly processing, where further recovery of materials and products is completed. At this stage some contaminated hazardous components and materials (e.g., batteries, lead-content materials) are collected and stored in a separate location for hazardous waste pick-up. Shredding and grinding of some materials is done for the purpose of densification or for security reasons. Distribution and outbound logistics are then completed by a number of internal and external representatives and vendors.

These processes and the operations associated with them are detailed in the next few sections.

2.2.1 Materials Suppliers (Inputs) for the AMRO Facility

The reverse logistics environment within AMRO causes purchasing, marketing, and customer service functions to overlap, and sometimes have unclear relationships. This makes the discipline of supply chain management even more critical, where it is assumed that supplier-customer relationships are core activities concerning both internal and external operational functions.

The AMRO facility has corporate (internal) and external suppliers of material. The internal materials supply includes equipment and materials from other groups and divisions within Compaq. Internal suppliers of material to AMRO include the manufacturing plants, physical plants, and any other facilities owned or operated by Compaq. These require a service-level agreement or internal contract. Some of the external suppliers include organizations that have various Compaq and brand-name electronic components that they no longer require, are obsolete, or are scraps. Typically, these external suppliers have a service-type agreement with Compaq.

External suppliers range from public and private organizations, including multinational corporations and municipalities. These suppliers of material to the AMRO facility come from all stages within the supply chain. The original equipment manufacturers may deliver material direct from their own processes (scraps) or inventory (obsolete material). Brokers serve a collection role and as intermediaries for the disposal or resale of products. Assembler suppliers are similar to manufacturers in that scrap and obsolete material may be sent to the AMRO facility. And financial institutions and government agencies, the end users, dispose of their equipment, usually due to upgrading or equipment beyond repair.

An important part of the marketing strategy to increase the number of external suppliers and offer a more consistent supply is to offer computer assets recovery as part Compaq's service portfolio. They have titled their services area as a services "store," and offer the following as part of their computer asset recovery services (Computer Assets Recovery Services brochure [Compaq 1999a]):

- Receiving returns—Scheduling, collecting, and receiving equipment. Visual inspection, sorting, and categorization.
- Product routing—Dealing with sorted products according to customer criteria

- Materials warehousing—Secure storage and inventory control
- Distribution—Picking, packing, and shipping resold or reallocated products and materials
- Direct Purchasing—Set-price purchase of unwanted equipment
- Recycling/Disposal—Expert disassembly of products. Environmentally safe disposal of components
- Project Management—Coordination of the end-to-end flow and reporting
- Remarketing—Resale of equipment into secondary markets
- Remanufacturing—Refurbishment of equipment, either for return to customer or for re-use
- Parts Recovery—Disassembly in order to sell recovered components

Whereas the first eight of these services are charged on a per unit basis, the last three service functions (remarketing, remanufacturing, and parts recovery) are priced on a revenue-share basis.

Suppliers of the raw materials can also be customers. Sometimes it is difficult to determine if a supplier would be considered an AMRO customer. It usually depends on whether there is an inflow or outflow of income from this supplier. That is, some suppliers of material get paid if the value of the material that arrives at the facility is capable of generating more revenue than transportation and processing costs. Either way, Compaq would typically view any supplier providing material to AMRO as a customer too—expanding the total quality management concept of everyone in a system serving as a customer.

2.2.2 Managing Supplier Accounts

The various customer/supplier relationships are managed by account managers, who conduct price and cost estimates on incoming and outgoing materials. The estimation and forecasting of the potential value of the materials is critical to determining the expected revenue and proceeds of the various deliveries. It is not a trivial task—requiring observation and constant updating of commodity market estimates (e.g., for plastics) to accurately forecast the amount of useable material within products. As well, the account managers require astute negotiation skills and detailed knowledge of product engineering and legal and environmental issues. Tools to aid them in this process are currently in great need. Guitini (1998) and Krupp (1992) discuss the difficulties in forecasting for remanufacturing environments, especially those with relatively immature reverse logistics. These tools need to go beyond those that purchasers and marketers have as information. It also requires greater integration among the various functions (engineering, marketing, purchasing, and logistics) to arrive at the most effective forecasts and accurate prices for material. Inter-organizational informational linkages would also benefit the operations within this function. The development and linkage of enterprise-wide information systems, such as enterprise resource planning (ERP) systems, could greatly benefit AMRO operations by accruing data from various functions internal and external to the organization. The linkage of supply chain management, ERP, and Internet systems, would provide a strong potential for more accurate pricing, forecasting, estimation, delivery, and overall project management of computer assets recovery.

Within the supplier relationship function, AMRO must identify where to find materials suppliers by evaluating internal and external customer markets. A bidding process for the supplies may then ensue (with external customers). The range of bids and the bidding process can go from simple bids on a single form, to long-term partnerships and alliances, which require a prolonged and detailed bid process with significant contractual obligations. This process requires estimation of revenue and the value of products, with the prices and costs negotiated on in the larger deals. Smaller bids are sourced from a standardized price list sent to potential suppliers of materials.

Once a supplier is found and a partnership is formed, the delivery of supplies needs to be managed. Aggregate scheduling and planning is difficult in this environment, since both the variability and uncertainty is substantial. With little advance warning and varying estimates of potential deliveries, only a few days notice is usually provided. There is some electronic tracking and monitoring of the material as it is delivered, but it only occurs when actual delivery takes place. Part of the reason for this type of situation is the difficulty in (or lack of) forecasting when, what, and why materials will be delivered to AMRO.

2.2.3 Disassembly and Processing of Materials

Materials for disassembly are trucked in from various domestic U.S. locations. Dock-workers (inbound logistics) are responsible for inspecting 95 percent of the inbound materials to determine if they are to be re-used (and thus stored in the warehouse as inventory), held for further evaluation and inspection, or sent for disassembly. Workers within the disassembly and final sorting areas inspect the remainder of the inbound material.

The initial layout of the disassembly area, called the property disposition center (PDC), was not very efficient. Changes were made to focus the layout on material cells (where a number of operations were carried out in one cell, depending on the product or material) rather than a single flow-through system where all operations are carried out on one product in one line. The PDC area layout underwent a continuous improvement process to help make disassembly operations as efficient as possible. The PDC, like most disassembly facilities, is very labor-intensive (much of it low-skilled) and heavily reliant on temporary personnel.

The material that arrives for disassembly is categorized into 89 specific types. As mentioned above, each of these materials is grouped into a sorting line. Plastics, electronic, metals, and glass are the major categories, and thus the lines are formed along these classifications. There are a number of parts that contain hazardous materials, such as nickel-cadmium batteries, lithium batteries, waste ink, and mercury. Therefore, disassembly requires meticulous attention to avoid misidentification. The facility needs to make sure that it reduces potential liabilities by minimizing misclassification. This issue becomes even more critical due to organizational restructuring within EHS and ISO 14000 standards, which support the use of environmental management at the source (similar to quality at the source) for managing environmentally sensitive processes. Thus, source reduction, a major component of the quality and environmental operations at the facility, helps to lessen the need for EHS personnel to oversee operations. Yet, the complicating factor at this facility is that most of the labor on the disassembly floor is contract or temporary employees.

Few permanent workers are part of the employee pool on the facility's shop floor. The difficulty with temporary personnel is that they need relatively quick and easy-to-understand training for them to be empowered. Their major duties involve the disassembly and sorting of old or returned electronics equipment. One of the issues in the sorting process was to minimize the amount of hazardous material and scrap waste that left the facility. Thus, processes were reengineered (improved layouts and flows) to make the process as simple as possible. Once again, a grouped, cellular layout allowed employees to focus only on the process requirements of a single type of material or product. Training had to focus more on understanding the various materials and components that flowed through this process, rather than on the steps in the process. To help complete this training, such that a zero amount of hazardous material left the facility, the workers were provided with visual aids and associated on-the-job training for the major parts. These visual aids, located not far from the work centers (in the relatively large facility), are composed of information boards with examples of the actual materials and parts clearly labeled and colored to allow for proper discrimination. The visual aids clearly show which parts and components come from certain product types and what to look for in terms of valuable or hazardous material. Especially critical is the discernment between hazardous and non-hazardous material. Even with temporary workers, downsizing of staff, and re-engineering of processes, the result of these improvements was extraordinary in virtually eliminating the need for hazardous material shipping or accidental loss.

The most experienced and skilled workers are assigned to lines that have the most complexity and hazardous material associated with the product. Many times, with a large variety of incoming materials, workers may come across unfamiliar products. These are stored in a non-conforming area, and an environmental and quality management team is called in to evaluate the materials.

Simplicity in process design not only aids low-skilled employees complete their job, but also helps in saving costs by requiring less time and activity for disassembly. This simplicity can be greatly enhanced with "design for disassembly" programs, some of which the operations managers within AMRO are involved with.

The activities within AMRO's disassembly operations, with respect to environmental management systems, are summarized in Table 3, which maps some of the major ISO 14000 steps and activities prevalent within the facility. Figure 2 shows the major operations within each of the disassembly lines, which includes disassembling the parts, sorting, decontaminating, and then grinding them. Some materials require all these steps, while others go through a subset of the operations. For example, printed circuit boards with a number of reprogrammable computer chips that are re-used in toys do not require grinding.

Table 3: Summary of environmental management system steps within AMRO.

Generic process step	**Process activity**
Process for identifying environmental aspects & impacts	Formally assessed by corporate EHS group annually.
Objective	Minimize potential liabilities.
Roles and responsibilities	Included in everyone's job plan and reviewed quarterly.
Enhancement of awareness and competency by training	Three levels of training: orientation training, on the job training, and annual awareness training.
Process for corrective actions	Corporate EHS audits twice a year.
Process for EMS review by top management	Top management reviews the results of the audit. Number of non-conformance and reduction in potential liabilities are considered effective measurements.
Results	Implementing continuous source reduction program though operational improvement.
Formalized TQM/JIT programs?	ISO 9000

2.2.4 Outgoing Materials Management

Similar to the inbound materials suppliers that are also customers, the definition of "customers" for the outgoing processed materials from AMRO overlaps with the traditional definition of vendors. Table 4 presents a summary of the various vendors for the AMRO facility. Again, similar to the relationships with the inflow materials suppliers, many of these outflow materials vendors or customers may either charge AMRO for removing the material or pay AMRO for these same materials. That is, AMRO may either receive revenue or increase costs, depending on the value of the material that is to be shipped out. Most of these vendors are approved and certified by Compaq corporate headquarters, and asked to submit bids on a weekly or biweekly basis. The outflow material suppliers and vendor base has shifted over the life of the AMRO facility. Initially, its major processes focused on the complete destruction and disposal of the equipment and components that flowed through it, because of the proprietary nature of the material, primarily owned by Compaq. As can be seen in Figure 3, "shearing" of printed circuit boards still occurs for security reasons in maintaining the confidentiality of proprietary technology. The chips and their technology are destroyed on purpose so that intellectual property and patented technology is not as easily accessible to competitors. For example, some of these boards and systems include patented application-specific integrated circuits (ASICS). This shearing is not only done to Compaq material, but also on the materials from companies that have contracted with Compaq for demanufacturing.

During the early recovery periods of AMRO, the products were smaller and lighter, with less precious metal content, making reclamation a less profitable enterprise. What wasn't recycled or reintroduced into other products was disposed of.

With the focus on only Compaq equipment and components, the remarketers and brokers were less interested in the outflow of materials. A new strategy that was introduced to disassemble and process brand-name components and equipment provided opportunities to increase the numbers of brokers and remarketers. Now that computer components and products have become more mature within the product life cycle, the amount of components and parts that have proprietary characteristics has decreased. This mature product environment allows for easier resale and marketing of materials. Yet, even in a multi-vendor environment, many suppliers of material to AMRO require full destruction of their products, thus limiting the markets for re-usability and remanufacturing.

Table 4: Categories and types of "vendors" for outgoing material—AMRO Resource Recovery Facility.

Recyclers (Basic materials)	**Re-users and Remanufacturers (Parts and components)**
Copper smelters	Brokers
Hazardous waste disposal	Secondary parts suppliers
Precious metal recovery	Compaq service parts stream
Plastics grinders	Repairable or directly re-usable material
Scrap paper	Toy manufacturer
Scrap metal recycler	

Reintroducing re-usable components to the market is slowed for competitive reasons. Thus, we see that maturing product life cycles (and strategic proprietary reasons) have an impact on the types of operations that an asset recovery facility may pursue. The types of process requirements are typically included in any contractual agreements.

3. BUILDING LINKS IN A GREEN SUPPLY CHAIN

Over the years, a few "test" relationships have existed between AMRO and its outflow materials vendors, and a number of articles have been written about them (Ashley 1993; Bergstrom 1993; Fiksel 1995; Riggle 1993; *Wall Street Journal* 1991).

One of the best-known examples of customer-supplier cooperation in a green supply chain context was a demonstration project operated jointly by AMRO, General Electric Plastics, and Nailite Inc. of Miami, Florida. Obsolete computer products were recovered from customers and dismantled in the AMRO facility. The computer housings were shredded, the plastic was reprocessed and then delivered to Nailite, where it was used to manufacture roofing products for McDonald's restaurants. The final conclusion of this pilot study was that it wasn't economically viable due to limited demand from an economies-of-scale perspective. A second project focused on the glass material from cathode ray tubes (CRT). This project was mounted in response to the U.S. EPA's efforts to more stringently limit the amount of lead allowed into the environment. Envirocycle joined with AMRO to develop a process for the leaded glass. Envirocycle approached Corning Asahi Video (a partnership between Corning and Asahi Glass in Japan) about possibly testing and ultimately using the recycled glass to manufacture new monitors and televisions. The same conclusion was reached as in the previous example—not economically viable. Yet, once again, the economics of the whole supply chain has yet to yield a scheme for profitable multi-organization materials remanufacturing and recycling on a large scale. Yet, strategic efforts do continue. Example possibilities of end items for CRTs and plastic coverings have been studied by a number of agencies. A distillation and summary of some of these items (beyond those described in this chapter) appear in Table 5.

4. SUMMARY AND CONCLUSIONS

The investigation of operations within a critical reverse logistics demanufacturer or disassembler can aid in understanding the amount of effort and requirements necessary for managing such an organization. With the increasing emphasis on greening operations, supply chains, and industrial systems, the investigation of these facilities can provide substantial insights for furthering the efficient operations of reverse logistics channels. A number of the practices conducted in a facility that demanufactures for a major computer and electronics organization (Compaq) were discussed. The reasons for demanufacturing computers and electronics components range from requirements of inventory management, proprietary product management, and waste reduction—and at the same time—the need top remain a profit-oriented service operation for customers.

Managing such a facility from various supply chain perspectives (inbound logistics, processing of materials, outbound logistics) can prove challenging in the emerging demanufacturing environment. A number of characteristics are very similar to those of a traditional manufacturer (e.g., supplier/customer relations, need for planning and control tools, flow lines), but there are also issues that make this environment unique.

In the case of Compaq's facility there are a number of lessons learned and issues that emerged.

- The facility's operations were very manually oriented. Little, if any, automation of controls systems existed. Clearly, this issue resulted in a need for manual labor in the plant. To keep expenses low and labor work levels relatively flexible, Compaq relied on temporary workers. This issue meant that they had to make work processes easy to learn and easy to understand. This issue was addressed by training and simplifications (process reengineering) of the process to bring these workers up to speed more quickly.
- Efficiencies in the flow of operations were also improved by reviewing and improving on the layout of the plant. Instead of a single line disassembling

material, parallel lines or cells that targeted specific assembly components and product types were developed. This may have led to some redundancies, but training was easier so that workers could focus on specific attributes of the products.
- To help increase the markets for their products, Compaq formed a number of partnerships to investigate possible uses and re-uses for materials that were plentiful and costly to dispose of, such as CRTs and plastics. The successes of these projects were still limited, but the knowledge from working with vendors and suppliers was helpful in furthering the possibilities of re-use.

Table 5: Sample of CRT glass and electronics plastic applications—existing and potential.

	Recycling category	**Application use**	**Material**	**Market development**
Cathode ray tubes	Re-use	Refurbish & resale	Variable	Available for some equipment
	Glass-to-glass closed loop	CRT glass manufacturing	Panel and funnel glass	Available
	Glass-to-glass open loop	Decorative tile	Co-mingled CRT glass	Demonstration phase
	Lead re-utilization	X-ray shielding products	Co-mingled CRT glass	Research and development phase
	Glass aggregate	Smelters: lead and copper	Co-mingled CRT glass or whole monitors for copper smelters	Available
	Export	Refurbish and resale: glass bottles for pesticides	Variable	Available
Plastics	Recycling—open loop	Asphalt products	Granulated plastics	Available
	Recycling—open loop	Plastic lumber	Granulated resins, some contamination allowable.	Available
	Recycling—open loop	Filler for thick-walled products	Loose plastic housings with metals removed.	Available

Source: Adapted from Dillon (1998).

Some of the future challenges include maintaining the profitability of the organization and control of the variability of the products (material inputs) arriving at the facility. As the AMRO facility went through the learning curve, it still had difficulty in determining and forecasting inbound materials. Forecasting models based on historical data or those more causally related to economic data will greatly benefit demanufacturing organizations in determining possible future return rates.

Generally, at a macro-level, industry perspective (issues also faced at AMRO), more research and investigation into this environment may provide managerial paradigms that can be useful to all organizations seeking to enter this type of market or industry. Some example areas of operational measures that could use further investigation include the development of forecasting and control tools for disassembly planning and management. Some work in this area has been completed from an optimization research perspective, specifically dealing with disassembly modeling (e.g., Das et al. 2000; Schuur and Krikke 1999; Zha and Lim 2000). A Delphi study investigating what the future holds for electronics disassembly (Boks and Tempelman 1998) concluded that (1) the obstacles of future disassembly and recycling technology are of an economic, rather than a technological nature; (2) automatic disassembly will probably gain importance in the next 25 years, but only for specific product categories; and (3) sorting and separation techniques will become increasingly sophisticated. Guide (2000) has identified additional practice and research needs for remanufacturing from an operations perspective.

ISO 14000 implications in this situation were quite evident. One of the major selling points (as a service to Compaq customers) of demanufacturing is its environmental soundness. The company viewed ISO 14000 as fitting well within the vision and strategy for the AMRO facility. Another reason for ISO 14000 certification at this plant was that it had previously achieved ISO 9000 certification for its quality system. Management viewed the additional step for ISO 14000 certification as a minor cost with respect to its benefits. The resulting impact on the organization was that it performed well in terms of minimizing its hazardous and general waste output because of the response to the various certification steps. Probably the most profound operational relationship was the diffusion of environmental responsibilities to shop floor workers through training and empowerment.

Another major issue is the required inter-organizational practices associated with reverse logistics relationships. For example, treating the supplier as a customer is one of the immediate practices that may help most organizations. This type of interaction is currently evident with "best practices" associated with those organizations that seek improvements in their relationships within a supply chain and have implemented inter-organizational total quality management (TQM) programs (Sarkis 1995b). Lessons learned in this environment may be particularly useful in managing suppliers in regular forward logistics environments. Here, the profitability associated with processing returned products has implications for both the supplier and the operations facility. The sharing of benefits and revenues is something that can be investigated for forward logistics relationships, as well.

Overall, from a supply chain perspective, the development of channels for products and materials will also be necessary. That is, an economically viable reverse logistics and demanufacturing supply chain is not possible without a critical mass of returnable products and materials that can be a constant supply through efficient channels. Modeling and managing the whole closed-loop supply chain environment to determine the most economical approaches also needs to be investigated. Some initial research seeking to optimize the operational efficiency in this environment has been completed (Jayaraman et al. 1999), but optimization is only part of the answer. Consumer acceptance and regulatory and legal issues also need to be managed. One such case involved counter-lawsuits between Compaq and Packard Bell, which charged each other with the use of refurbished (instead of new) products (Ouellette and Vijayan 1995). Which leads to the issue of what is considered new and what is refurbished, as well as the quality associated with such products. Without consumer demand and acceptance of refurbished products, the remanufacture and re-use of these systems and their components will be seriously limited.

There have been some efforts at Compaq to further the relationships between initial product design and the demanufacturing operations for the disassembly of these products. This issue becomes even more critical as public and government sectors continue to discuss and define extended product and producer responsibilities. Part of meeting these challenges is the need to have an integrated closed-loop communications process between product design and demanufacturing facilities (Grenchus et al. 1998). Design issues facing demanufacturing range from improving dismantling efficiencies, reducing mixed plastics designs, and the reduction or elimination of hazardous materials (e.g., batteries) in the design of products (Grenchus et al. 1997; Das and Matthew 1999). Further development and investigation of a closer link between these initial development and final disposition phases is also needed. To aid in this process at Compaq, account managers that managed the material supply chain came from technical and marketing functions. This characteristic allows them to more effectively communicate with product design and development departments, especially when the focus is on design for disassembly issues.

This is only a summary of the many issues revealed in the case study. More research and study is always warranted for this area due to its relative immaturity. Operational improvements exist around every corner, and many of these improvements can only contribute to the organizational and environmental bottom-line.

5. EPILOGUE

As of the winter of 2000, Compaq decided to close down the AMRO facility and outsource the demanufacturing function to DMC Incorporated, a major regional electronics demanufacturing company. Even though the facility was profitable, it did not meet Compaq's strategic vision, and due to the world-wide high technology economic downturn, they decided to focus internal resources elsewhere. Yet, this product recovery and demanufacturing service is still offered to customers. The strategic vision included a merger with Hewlett Packard, Inc., which was announced in September 2001.

REFERENCES

Ashley, S. 1993. Designing for the environment. *Mechanical Engineering* 115:52–55.

Bergstrom, R.Y. 1993. An annotated essay: Environmental affairs. *Production* 105:36–41.

Bloemhuf-Ruwaard, J. M., P. van Beck, L. Hordijk and L. N. van Wassenhove. 1995. Interactions between operational and environmental management. *European Journal of Operational Research* 85:229–243.

Boks, C. and E. Tempelman. 1998. Future disassembly and recycling technology: Results of a Delphi study. *Futures* 30:425–442.

Carter, C. R. and L. M. Ellram. 1998. Reverse logistics: A review of the literature and framework for future investigation. *Journal of Business Logistics* 19:85–102.

Compaq. 1999a. Computer store: Computer asset recovery. Compaq brochure CS-OC-764-01/99 08, United Kingdom.

———. 1999b. Computer asset recovery service: Responsible management for your obsolete computer equipment. Doc. No. 5994/1099, Compaq Computer, Australia.

Das, S. and S. Matthew. 1999. Characterization of material outputs from an electronics demanufacturing facility. 1999 IEEE Proceedings of International Symposium on Electronics and Environment, Danvers, Massachusetts, 251–256.

Das, S. K., P. Yedlarajiah and R. Narendra. 2000. An approach for estimating the end-of-life product disassembly effort and cost. *International Journal of Production Research* 38:657–674.

Dillon, P. S. 1998. Potential markets for CRTs and plastics from electronics demanufacturing: An initial scoping report, technical report 6. Chelsea Center for Recycling and Economic Development Technical Research Program, University of Massachusetts, Lowell.

Grenchus, E., R. Keene and C. Nobs. 1997. Demanufacturing of information technology equipment. 1997 IEEE Proceedings of International Symposium on Electronics and Environment, 5–7 May, at San Francisco, CA.

Grenchus, E., R. Keene, C. Nobs, A. Brinkley, J. R. Kirby, D. Pitts and I. Wadehra. 1998. Linking demanufacturing operations with product DFE initiatives. 1998 IEEE Proceedings of International Symposium on Electronics and Environment, 4–6 May, at Oakbrook, IL, 270–275.

Guide, V. D. R. 2000. Production planning and control for remanufacturing: Industry practice and research needs. *Journal of Operations Management* 18:467–484.

Guitini, R. 1998. Forecasting in remanufacturing. Fifth International Conference on Environmentally Conscious Design and Manufacturing, 16 and 17 June, at Rochester, NY.

Fiksel, J. 1995. How to green your supply chain. *Environment Today* 6:29–30.

Fleischmann, M., J. Bloemhof-Ruwaard, R. Dekker, E. van der Laan, J. A. van Nunen and L. N. Van Wassenhove. 1997. Quantitative models for reverse logistics: A review. *European Journal of Operational Research* 103:1–17.

Jayaraman, V., V. D. R. Guide and R. Srivastava. 1999. A closed-loop logistics model for remanufacturing. *The Journal of the Operational Research Society* 50:497–508.

Krupp, J. A. G. 1992. Core obsolescence forecasting in remanufacturing. *Production & Inventory Management Journal* 33:12–17.

Lamming, R. and J. Hampson. 1996. The environment as a supply chain management issue. *British Journal of Management* 7:45–62.

Narasimhan, R. and J. R. Carter. 1998. *Environmental supply chain management*. Tempe, AZ: The Center for Advanced Purchasing Studies, Arizona State University.

Ouellette, T. and J. Vijayan. 1995. Packard Bell, Compaq head for showdown. *Computerworld* 29:32.

Pepi, J. 1998. University of Massachusetts Amherst scrap electronics processing, technical report 7. Chelsea Center for Recycling and Economic Development Technical Research Program, University of Massachusetts, Lowell.

Pohlen, T. L. and M. T. Farris. 1992. Reverse logistics in plastics recycling. *International Journal of Physical Distribution & Logistics Management 22:35–47.*

Porter, M. E. 1985. *Competitive advantage: Creating and sustaining superior performance.* New York: The Free Press.

Riggle, D. 1993. Component recycling for old computers. *Biocycle* 34:67–69.

Sarkis, J. 1995a. Reverse logistics, recycling and the product life cycle. Proceedings of GEMI '95 Conference: Environment and Sustainable Development, May, at Washington, D.C.: 257–264.

———. 1995b. Supply chain management and environmentally conscious design and manufacturing. *International Journal of Environmentally Conscious Design and Manufacturing* 4:43–52.

———. 2001. How green is my supply chain. Working paper. Worcester, MA: Clark University.

Schuur, P. C. A. and H. R. Krikke. 1999. Business case Roteb: Recovery strategies for monitors. *Computers & Industrial Engineering* 36:739–758.

Shrivastava, P. 1995. Ecocentric management for a risk society. *Academy of Management Review* 20:131.

Turton, R. 2000. EOL recycling technology assessment. <http://www.electronicsrecycling.net/menu2/industry/projects/htms/prjWVUturton.htm> (8 January 2003).

Wall Street Journal. 1991. GE, firms form venture to boost plastic recycling. *Wall Street Journal,* 18 June, B5.

Walton, S. V., R. B. Handfield and S. A. Melnyk. 1998. The green supply chain: Integrating suppliers into environmental management processes. *Supply Chain Management* 34:2–11.

Zha, X. F. and S. Y. E. Lim. 2000. Assembly/disassembly task planning and simulation using expert Petri nets. *International Journal of Production Research* 38:3639–3676.

Chapter 13

MANAGING PCS THROUGH POLICY: REVIEW AND WAYS TO EXTEND LIFESPAN

Ruediger Kuehr
United Nations University Zero Emissions Forum, Germany

1. INTRODUCTION

On May 12, 1941 a young German named Konrad Zuse presented his latest development to a small, select group of people. Unnoticed by the public, the student of civil engineering at Berlin's Technical University had succeeded in making his dream a reality. It was called the "Z3," the world's first functioning programmable calculator, and it marked the start of a new era. The operational target that Zuse accomplished was to replace the dull necessity of manual calculations with a machine. The mechanical contraption, made of 40,000 components and 2,500 telephone relays, was able to save 64 numbers in its memory and execute basic arithmetic operations (Konrad Zuse Internet Archive 2001).

With the first technical development of a computer 62 years ago, the basis for a revolution was laid, which today influences the daily lives of people around the world. With ongoing miniaturisation, in combination with enormous increases in performance and quality, computers have become a tool accessible to millions of people, many who proudly bought their first computer in the late 1980s or early 1990s and have upgraded their systems several times since. Furthermore, the continuous spiral of shrinking prices has led to a virtual explosion in personal computer sales.

Consequently, the personal computer (PC) has almost become ubiquitous—indispensable in many sectors like air and space traffic control, medicine, schools, the mass-production of many daily goods, military, and even in private households. The benefits of the computer boom have unquestionably been widespread, but it has not been entirely positive. The enormous material and energy flows required for the manufacture and use of computers—with their short life cycles and the growing mountains of computer waste left behind—cast a long shadow, which is extensively described in the other chapters of this book.

Computers and the Environment: Understanding and Managing Their Impacts
Edited by Ruediger Kuehr and Eric Williams, pages 253–278.

The analysis of policies for extending the PC's life cycle is confronted with three fundamental problems of political science, which can be summarised under the headings of complexity, dynamism, and possibilities. Complexity describes the interrelationships between different actors and the interdependencies between norms, values, and behavioural patterns. Dynamism describes the quality of interactions among those relationships, the interdependencies of the various elements, and the elements themselves caused by continuous change. This complexity of problems and continuous change lead to an ever-growing constellation of problems to be solved. Political science aims to develop concurrence between the "real" world and our worldviews, making it necessary to abstract. Thus, a certain part of "reality" has to be identified, selected, and analysed from a specific perspective—here, the path towards a more sustainable PC.

Consequently, the basic data on the information technology (IT) sector already provided throughout this volume, focusing predominately on the PC sub-sector, helps to impart information that's essential to a comparative assessment of the overall balance and performance of related policies in three countries—Japan, Germany/the European Union, and the United States.

As described, the economic and social significance of the IT sector and its high growth rates are important indicators of the need for political and legislative action, as are the enormous material flows within this sector that are resulting in growing mountains of waste. Thus, the statistical data provided an essential platform for identifying the necessity of getting the PC problem onto the political agenda. The first part of this chapter shows the different developments coming out of rather similar initial situations by comparing governmental policymaking on the IT sector in Germany/European Union, Japan, and the United States. Wherever possible, I try to briefly explain the occurrence of these differences, but the given framework of this analysis neither allows a deep comparative excursion into the complex system nor would it be appropriate with the focus of this book. On the contrary, the goal of this study is to present different policies on how to deal with the PC's life cycle and to come to an initial assessment of those, based on the results of the other chapters.

At the same time, it is undeniable that the responsibilities and actions of industry as well as the initiatives of non-governmental organisations (NGOs) are influential. Their leadership in forging progress has partially been used as an argument for the common "wait-and-see" attitude often taken by national governments, which usually take a more passive role because of the engagement of others. The influence of industry and NGOs on the policymaking process is thus presented and compared in the second part of this chapter.

An in-depth evaluation here of the political and legal measures taken is impossible. On the one hand, many of the legal steps have not yet been implemented, so there is a lack of practical experience, and some aspects of effects cannot be predicted. But at the same time, an initial evaluation should definitely be made in terms of the "sustainability" paradigm—adopting a holistic perspective that should include many more details (e.g., on process engineering, total material flow, cost-benefit analysis, social acceptance, etc.). For this reason I decided to refrain from offering comprehensive final conclusions, but to explicitly tackle a concrete assessment of the political and legal decisions taken on the issues. Out of this I develop suggestions on how to proceed to effectively manage the impact of personal computers towards more sustainable development in the future.

2. LAWS AND POLICIES

The social, economic, and ecological significance of the IT sector, especially of PCs, is indisputable, and was extensively described in the introduction and the other chapters of this book. In the face of growing mountains of PC waste, the question arises: Does it appear that existing policies and laws are appropriate and efficient in controlling these enormous material flows, while taking the sustainability paradigm into account?

2.1 Germany

2.1.1 Laws

The Law for the Promotion of Closed Substance Cycle Waste Management (KrW),[1] which came into force 7 October 1996, lays down the principals to secure the realisation of an ecological cycle economy of resources by means of a range of product-specific regulations and directives. Consequently, the KrW also forms the basis for the treatment of electrical and electronic waste in combination with the general directives on dealing with waste—the so-called TA Waste and TA Domestic Waste.[2] Following these edicts, untreated electrical and electronic waste must be disposed of separate from the regular waste stream and can no longer be disposed of in landfills or other waste dumps. Detailed regulations are prescribed in the municipal statutes of cities and local authorities (UBA 1996, 15).

An essential element of the KrW is product responsibility,[3] as laid down in Paragraphs 22 to 25. Producers and distributors are now responsible for the re-introduction of these products into the production process, as well as for the ecologically sound disposal of those components which cannot be re-used. The actual implementation of this type of product responsibility in the individual sectors of

[1] In German, KrW stands for Kreislaufwirtschafts- und Abfallgesetz, or law for the prevention, utilization, and disposal of waste, or law supporting cycle economy and ensuring the environmentally sound disposal of waste.

[2] In German, TA stands for Technische Anleitung, or technical directive.

[3] Use of the "responsibility principle" is intended to end the current situation, whereby the industry manufactures the products but the municipal authorities have to dispose of the waste generated at the cost of the general public.

industry requires further legal steps by the federal government. According to Paragraph 24, the requirements and responsibilities for the assumption of costs for the recycling/disposal of the products by the producer or distributor are starting to be laid down in the legal directives passed by the federal government.

The substantial resistance of the German automobile industry against the "old car regulation" (which forces the industry to take back old cars free-of-charge), as well as the rebellion of another German industry against the deposit on beverage cans to guarantee high return quotas for recycling (which came into force 1 January 2003), substantially clarifies the problems with the implementation of single national regulations. This resistance is supported by the still relatively low prices of natural resources on the world market, which make it more attractive to use virgin material than recycled materials. Additionally, the investments and costs required for the take-back systems need to be recognized.

Even for small and medium-size companies, which characterize the German industrial landscape, these types of regulations seem to form a substantial burden. And since prices are still the most important indicator for sellers and consumers, Germany's domestic industry loses out from the short-term economic point-of-view against importers from other countries, who can produce cheaper products because of more lax environmental regulations.

On the other hand, Germany's present taxation system gives impetus to a further increase of the mountains of PC waste, and thus contradicts may other approaches, by allowing business and private consumers to totally depreciate the cost of a new PC within a three-year period. Consequently, after exactly three years they buy a new PC (brand-new usually) and dispose of their "obsolete" equipment in a market where interest, for the most part, does not yet exist. Buying a used PC to replace the 100-percent depreciated computer is usually out of the question, since it might be too inexpensive and thus not worth it in a new round of depreciation, besides not being up-to-date. Consequently, the German tax system blatantly only supports new PCs, whereas re-use and refurbishment are not reflected.

2.1.2 Policy Initiatives

Just before the KrW was passed in 1991, a first draft of a directive for the environmentally sound disposal of electrical and electronic equipment (*Elektronikschrottverordnung* in German) was tabled by the German federal minister of the environment of the time, Klaus Töpfer (now executive director of United Nations Environment Programme). A hearing of the parties concerned took place, as planned in the Waste Act, and the amended draft was completed. Although it was initially intended that the directive be implemented in 1994, then in 1996, no binding regulation for the IT sector has yet been approved (Bundesanstalt für Arbeitsschutz und Arbeitsmedizin 1997, 7).[4]

The draft directive on electrical and electronic equipment waste touches on a broad range of products. Consequently, it is practically impossible to set similar regulations for different categories (Thomsen 1997, 255). This led to the drafting of a directive for IT products in 1995 by the working group CYCLE, in agreement with the German Federation of the Electro-Technique and Electronic Industry (ZVEI). It prescribes the take-back and recycling of IT products by the producers on a voluntary basis (UBA 1996; Ebeling 1996). With direct reference to this, the German Federal Ministry of the Environment (BMU) presented the amended draft, called the Directive on Waste IT Appliances (ITVO),[5] which was concerned with the approximately 110,000 tonnes of computers, monitors, scanners, typewriters, photocopiers, fax machines, etc., that already made up about 10 percent of the total amount of electronic waste in 1995[6] (but is very likely to be much higher today) (UBA 1996, 17; Bundesministerium für Umwelt 2001). This draft was in line with the European guidelines on waste, with product responsibility as the fundamental principle and extending it to the disposal phase. Although the ITVO passed the German Bundestag (federal parliament) in 1998, it still has not been approved by the German Bundesrat[7] (state's federal body) (Bundesratsdrucksache 1998).

In their coalition agreement in 1998, in laying down the principles for the federal government, the political parties—the Social Democrats and the Greens—agreed to meaningful regulation of all aspects of electrical and electronic wastes (Grote 1999). All federal states governed by the Social Democrats, and other states as well, recommended considerable improvements to the ITVO. Thus, the original ITVO draft remained, but was extended to electrical and electronic equipment (EAV).[8] The Environmental Committee of the Bundesrat made the following suggestions for modifications in June 1999:

- The EAV should also include so-called white goods (refrigerators, cookers, washing machines, etc.), brown goods (TVs, stereo systems, etc.), and other small appliances. Consequently, it points to a comprehensive directive on electrical and electronic wastes and thus for a majority of the 1.5 million tonnes of waste generated each year.

---

[4] The KrW took on the role of the original directive regarding electronic waste in many areas, but its in-depth effectiveness is questioned (Bundesanstalt für Arbeitsschutz und Arbeitsmedizin 1997, 7; compare also Donicht 1996).

[5] IT-Altgeräteverordnung in German.

[6] Of the 10 percent, 40 percent are "household appliances" and 20 percent are "leisure appliances" (Grote 1999, 90); two-thirds are from domestic households, the rest from the commercial sector.

[7] The Bundesrat is the link between the federal government and the states. It is the channel through which the states can participate in the legislation and administration of the Federal Republic and in EU affairs. The Bundesrat is involved in all federal legislative proceedings. Over half of all federal laws must receive its explicit approval before they can come into force. This gives it considerable influence over federal legislation.

[8] Elektro- und Elektronikaltgeräteverordung in German.

- Producers should also be obliged to take back the same amount of appliances they bring onto the market annually. This includes similar appliances to their own products and those that were manufactured before the directive came into force—the so-called historical wastes.
- Local authorities would be responsible for the collection of waste from private households, but would no longer be in charge of final disposal.

But the extension of the directive has not yet been decided on by the new federal government, although it has agreed on a far-reaching realisation of a directive on electrical and electronic waste in its coalition contract between the Social-Democrats and the Greens in 1998 (Grote 1999, 90). In the meantime, the draft guidelines drawn up by the European Union (E.U.) put the national legislative process on hold until 13 February 2003, when the final texts for the WEEE and RoHS directives came in publication format for the Official Journal of the European Union, demanding a transposition through the member states by 13 August 2004. Nevertheless, it is predictable that the entire German directive will be adjusted and rewritten along the new E.U. directives (Dworak and Kuhndt 2000). Additionally, the latest coalition contract (October 2002) of the Social Democrats and the Greens, which named sustainability as one of the three main objectives already in its title, no longer highlights or even mentions the need to channel the growing mountains of IT waste (Sozialdemokratische Partei Deutschlands 2003).

In May 2000, a joint motion of the Social Democrats and the Greens entitled "Roadmap for Sustainable Information Technology" passed the German Bundestag. Its purpose is to indicate core problems in this area, to identify technical challenges, and to demonstrate problem-solving means and approaches. Conceptually, close co-operation between the IT sector, the sciences, and politics should be strengthened to ensure a reliable frame of orientation.

2.2 European Union

2.2.1 Laws

Since 1975 the basic guidelines on waste have been aimed at standardising the legal regulations regarding waste in the member states of the E.U. The central goal of each regulation on waste disposal must be to protect both human health and the environment from any detrimental effects arising from the collection, transportation, handling, storage, and disposal of waste. Member states should take the appropriate measures to prevent or reduce the production of waste and encourage the recycling or correct disposal of any waste generated (Kunig 2003).

The guidelines on hazardous waste have served to standardise the legal regulations regarding the controlled management of hazardous waste in the E.U. member states since 1991.[9] Member states are responsible for monitoring each of their waste dumps for hazardous materials and identifying their source. The mixing of hazardous waste with other waste is not permitted, unless it is to enable the prevention, processing, and conversion of the hazardous waste (Europäische Kommission 1996).

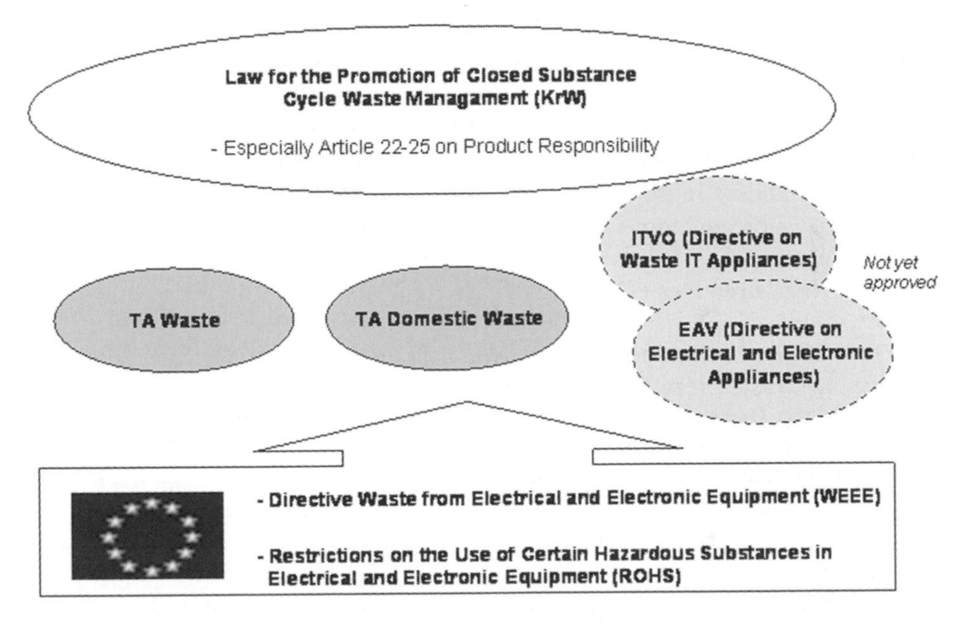

Figure 1: Germany's legal approaches on PC waste.

The E.U. guidelines described above form the only very general regulations on electrical and electronic wastes still in use today.[10] But after the adoption of the new WEEE and RoHS directives, the new recycling targets given should be met by 31 December 2006 in E.U. member states.

2.2.2 Policies

The European Community (E.C.) acknowledged the stream of waste from electrical and electronic equipment in 1994. By setting up an ad hoc project group named Waste from Electrical and Electronic Equipment, it initiated the preparation of proposals for E.U. action. The group submitted a recommendation in July 1995,[11] but the document

---

[9] Relevant materials in electrical and electronic waste are, for example, contaminated appliances, batteries, copper compounds, selenium compounds, lead compounds, PCBs, PCT, and compounds containing halogen (UBA 1996, 16).

[10] See, Council Guideline 91/689/EWG of 12 December 1991. This was later extended by Council Guideline 94/31 EG.

[11] Italian National Agency for New Technology, Energy, and the Environment. 1995. Priority waste streams: Waste from electrical and electronic equipment. Rome.

failed in its purpose of providing concrete proposals for E.C. action. As a consequence, the Commission froze movement on the initiatives. In the meantime, several E.U. member states took action toward developing legislative steps to channel the flow of waste electrical and electronic equipment (Welker 1996). As a result the Commission released a working paper on electrical and electronic waste in 1997. On 13 June 2000, after almost three years of intensive discussions, the E.U. Commission tabled drafts of the E.U. directive on waste from electrical and electronic equipment (WEEE) and the directive on the restriction of the use of certain hazardous substances in electrical and electronic equipment (RoHS). In certain parts, these directives go far beyond the German EAV draft. Additionally, this has developed a further dynamic during the period of amendment in the political arena and industrial sector, which undoubtedly points towards more ambitious regulation.[12]

2.2.3 New Directives

In late 2002, the European Parliament and the European Council adopted the Directive on Waste from Electrical and Electronic Equipment (WEEE)[13] and the Directive on the Restriction of the Use of Certain Hazardous Substances in Electrical and Electronic Equipment (RoHS).[14]

Following lengthy debates between the Council (member States) and the European Parliament Conciliation group (made up of 15 members of the European Parliament) there have been some revisions to the original proposal of the European Commission from May 2000 and some amendments to the Common Position text agreed in June 2001. Consequently, the final decision-making process, based on the proposal, lasted more than two years.

(1) Waste from electrical and electronic equipment (WEEE)

The WEEE Directive targets the regulation of the collection and disposal processes of electrical and electronic equipment waste.[15] Thus, consumers can return their appliances free-of-charge, with the producers responsible for the recycling or environmentally sound disposal of waste appliances. In the case of historical waste, a transitional period of five years is suggested before manufacturers are also obliged to take full responsibility for the recycling of these products.[16] The costs generated by appliances that were already in circulation at the time the directive came into force

[12] Christian Hey of the European Environment Board (EEB) in an interview in April 2001.

[13] Commission of the European Communities Proposal for a directive of the European Parliament and the Council on Waste Electrical and Electronic Equipment 2000/0158(COD), COM 2000 347 provisional, Brussels, 13 June 2000.

[14] Commission of the European Communities Proposal for a directive of the European Parliament and the Council on restriction of the use of certain hazardous substances in electrical and electronic equipment 2000/0159 (COD), COM 2000 347 provisional, Brussels, 13 June 2000.

[15] <http://164.36.253.20/sustainability/pdfs/finalweee.pdf> (13 March 2003).

[16] This applies to electric/electronic appliances sold before the directive came into force and for which the producer was unable to levy a recycling charge additional to the product price.

should be shared among all producers. Consequently, producers are being encouraged to develop appliances that are recyclable. By the end of 2006, manufacturers will be obliged to achieve ambitious recycling rates, with the minimum rate of recovery targeted at ranges between 50 and 80 percent, depending on the waste category. For the category that includes PCs, the recovery rate will be increased to 75 percent by weight of the appliances; and component, material, and substance re-use and recycling will be increased to a minimum of 75 percent.

The member states decide independently who will be responsible for the collection of waste appliances. In Germany, according to the agreement following the EAV, the local authorities will organise and finance collection from private households in all those cases which do not fall under the producers' responsibility principle. The retailer is furthermore obliged to take back a "similar" waste appliance when selling a product. The necessary collection and recycling structures should be in place by the year 2006. From 2006, a minimum rate should be achieved of an average of four kilograms of electrical/electronic waste per inhabitant per year collected separately from domestic households.

(2) Restriction on the use of hazardous substances (RoHS)

The restrictions on the use of certain hazardous substances in electrical and electronic equipment (RoHS) targets a standardization of the national laws of the E.C. members and tries to contribute to the environmentally sound recovery of disposal of waste from electrical and electronic equipment.[17] Consequently, in effect from 2008, the use of lead, mercury, cadmium, hexavalent chromium, polybrominated biphenyls (PBBs), and poly-brominated diphenyl ether (PBDE) must be substituted by other substances, with a few exceptions when the use of those materials is technically or scientifically unavoidable and/or health impacts are likely to outweigh environmental benefits (Commission of the European Communities 2000a). Legally, this proposed guideline is based on Article 95, "standardisation of the home market" in the Amsterdam Treaty,[18] and thus prevents national increases to large extent.

Previous chapters of this volume described, at least in some parts, the various viewpoints of the computer industry in regard to these new directives, which should be considered not only as challenges, but also as opportunities.[19]

---

[17] <http://164.36.253.20/sustainability/pdfs/finalrohs.pdf> (17 February 2003).

[18] The aim of this treaty, agreed to by the European Union's political leaders on 17 June 1997, is to create the political and institutional conditions to enable the E.U. to meet the challenges of the future, such as the rapid evolution of the international situation, the globalisation of the economy and its impact on jobs, the fight against terrorism, international crime and drug trafficking, ecological problems, and threats to public health.

[19] See, Chapter 5 by Höhn and Berkley (IBM), Chapter 6 by Podratzky (Fujitsu Siemens), and Chapter 4 by Hieronymi and Schneider (Hewlett-Packard).

2.3 Japan

2.3.1 Laws

The Waste Disposal and Refuse Collection Law (Law No. 137), passed by Japan's legislative organisations (the House of Representatives and the House of Councillors) in 1970, formed Japan's first legislative step against the growing problems of waste, in the face of the serious environmental crisis and after two decades of almost entire environmental ignorance during the 1950s and 1960s (Kühr 1996a, 1996b). Up until today, this law had been revised three times. The Resource Recovery and Recycling Promotion Law of 1991 was the first one to target the promotion of recycling, and furthermore, set recycling quotas. Legislation enacted in May 2000 aims at promoting a recycling-oriented society and spells out the basic principles behind the government's vision for conserving resources. Consumers and producers are required to recycle certain goods now that additional legislation for recycling home appliances and food is in effect. This new law aims at requiring everyone in the product cycle, from production to consumption to waste disposal, to make some contribution.

2.3.2 Policies

Although these laws provide a general framework for Japan's approach to handling waste from electrical and electronic equipment, the amendment of the Law for Promotion of Effective Utilisation of Resources by Japan's Ministry of International Trade and Industry (MITI) is the first one that explicitly mentions PCs. The main objective of this law is to pull away from the existing economic system, which is based on mass production and mass disposal, and to create a recycling-oriented economic system, for which certain measures have been adopted under the Resource Recovery and Recycling Promotion Law (RRRP). The new scheme created by the amendment of the Law for the Promotion of the Creation of a Recycle-Oriented Society provides the general framework.[20] It explicitly aims to introduce the so-called 3Rs strategy. Consequently, certain but very basic policy steps are prescribed to reduce the generation of industrial waste, including the generation of byproducts, certain measures applicable to products for re-use, and the utilisation of used products as raw materials for new production processes through recycling. The Waste Management Law and the Law for the Effective Utilisation of Resources are additional framework laws, but they are responsible for developing a more general scheme towards a recycling-oriented society. And another four laws that focus especially on recycling set regulations in accordance with the characteristics of individual products.

The Law for the Promotion of Utilization of Recycled Resources is designed to encourage companies and individual consumers to aid in efforts to recycle containers

---

[20] <http://www.meti.go.jp/policy/closed_loop/main_01.html> (4 April 2003) in Japanese.

and other products as raw materials (replacing fossil fuels) and to reduce the mountains of waste. It also provides a list of industrial sectors that must use recycled materials and a list of products that must adopt easy-to-recycle materials (Nikkei 2003). The government, under the guidance of the Ministry of Economics, Trade and Industry and the Ministry of Environment, revised the law so as to force manufactures (of PCs, for example) to re-use old components in making new ones. Earlier, the RRRP had already identified the PC as one example of a product that results in a huge volume of waste. Amendments to the RRRP Law took effect in April 2001, requiring manufacturers of 69 items of office equipment, including computers for company use, to recycle their used products. The aim is to recycle the rare metals used in motherboards and to make new computer cases with recycled plastic.

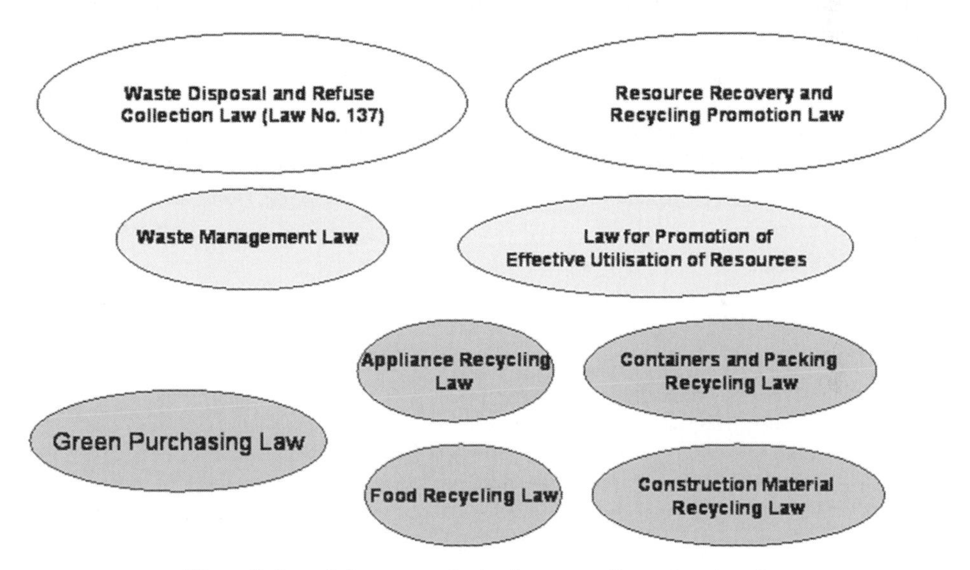

Figure 2: Japan's laws towards shaping a recycling-oriented society.

And from the fall of 2003, at the earliest, a recycling system for home-use personal computers will come into effect. Consumers will be required to pay recycling fees when they buy a new PC and when they dispose of their old one, which was purchased before the system's introduction. The respective fees will be included in the sale price of a new PC. In addition, PC manufacturers will be voluntarily required to collect used PCs from households, as well as businesses. Mitsubishi Electric Corp. and Samsung Japan Corp., in addition to Hitachi Ltd. and IBM Japan Ltd., agreed to conduct a joint PC recycling business. And under an agreement with 21 major computer manufacturers, Japan Post will collect used PCs at more than the 20,000 post offices all over Japan, or even pick them up from people's homes (*Japan Times Online* 2003). Please see Appendix 1 for further reading on PC-recycling policies in Japan.

2.4 United States

2.4.1 Laws

Enacted in 1976, the Resource Conservation and Recovery Act (RCRA) created regulations on solid and hazardous waste disposal in the United States. It significantly expanded the previous Solid Waste Disposal Act by adding a new subsection that deals specifically with hazardous waste. The RCRA gives the U.S. Environmental Protection Agency (EPA) the authority to control hazardous waste from "cradle-to-grave," which includes generation, transportation, treatment, storage, and disposal of hazardous wastes. If waste is not hazardous, the federal role is insignificant compared to that of the state (Patton-Hulce 1995, 263).

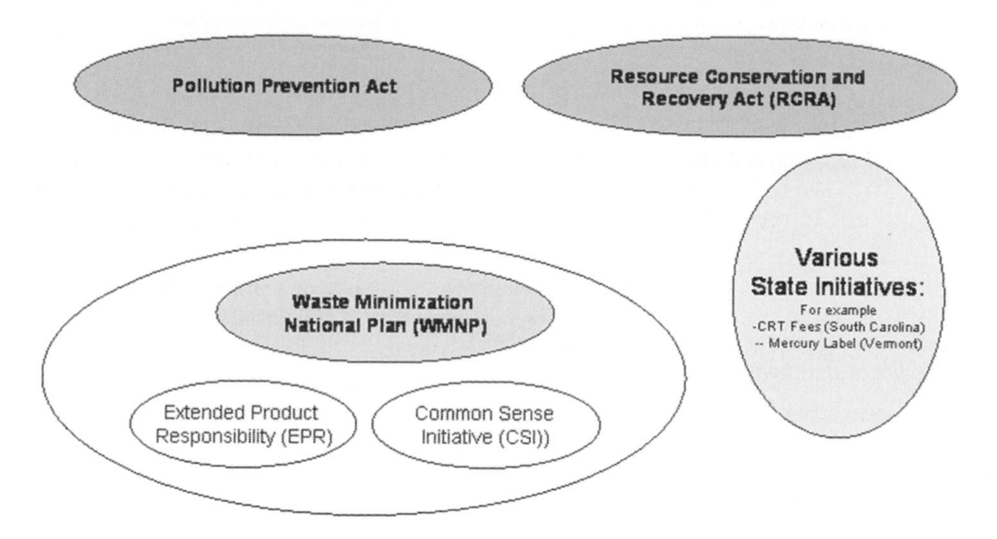

Figure 3: Legal policies on IT waste minimization in the United States.

As explained in previous chapters, the cathode ray tubes (CRTs) in colour monitors and televisions are hazardous to human health and the environment when discarded, because of the presence of lead in these products. Federal regulatory requirements applicable to handling these materials vary; state regulatory requirements, as well, can be different from federal regulatory requirements (EPA 2000, 4–5). Used computer monitors disposed of by households or sent for continued use (i.e., resold or donated) are not considered hazardous waste and not regulated under federal regulations. Federal laws only regulate waste from facilities that generate more than 100 kilograms of hazardous waste per month.

The Pollution Prevention Act (1990) focused industry, government, and public attention on reducing the amount of pollution through cost-effective changes in production, operations, and raw materials use.[21] Pollution prevention includes other practices that increase efficiency in the use of energy, water, or other natural resources and protects the resource base through conservation with practices such as recycling, source reduction, and sustainable agriculture.

Apart from soft laws and policy programmes, no legislative action on WEEE is envisaged soon at the federal level in the United States (Commission of the European Communities 2000b).

2.4.2 Policies

To encourage waste minimization nationwide, the EPA worked with representatives from numerous stakeholder organisations to develop the Waste Minimization National Plan (WMNP), which was released in November 1994. It focuses on reducing the generation and subsequent release into the environment of the most persistent, bio-accumulative, and toxic chemicals contained in hazardous waste. The WMNP established three goals:[22]

1. As a nation to reduce by 50 percent the presence of the most persistent, bio-accumulative, and toxic chemicals in hazardous waste by the year 2005;
2. To avoid transferring these chemicals across environmental media; and
3. To ensure that that these chemicals are reduced at their source whenever possible, or, when not possible, that they are recycled in an environmentally sound manner.

The EPA's Extended Product Responsibility (EPR) is a product-centred policy approach to environmental protection. It calls on those involved in the product life cycle—manufacturers, retailers, users, and disposers—to share responsibility for reducing the environmental impacts of their products (EPA 2000).[23] The U.S. EPA is actively facilitating coordination and collaboration among states, local governments, industry, and non-governmental organisations on these issues. The EPR program has supported projects in a number of product areas, ranging from electronics to carpets.

The EPA's Common Sense Initiative (CSI), launched in 1994, represented a new approach in creating policies and environmental management solutions that relate to industries. It focused on six industry sectors (automobiles, manufacturing, computers, electronics, metal finishing, and petroleum refining), and tried to help develop cleaner, cheaper, and smarter solutions for industry and taxpayers (EHC 1998, 28–29). Several initiatives related to the computer and electronics industry, for example, through (a) filling the data gap, (b) reducing regulatory barriers to CRT recycling, and (c) increasing public awareness.

Despite the large volumes of electronic equipment in U.S. households, residential collection is not currently under consideration. There are, however, some examples of progress. Massachusetts recently banned the disposal of CRTs in its municipal waste

---

[21] <http://www.epa.gov/region5/defs/html/ppa.htm> (04 April 2003).
[22] <http://www.epa.gov./epaaoswer/hazwaste/minimise> (03 April 2002).
[23] EPA online <http://www.epa.gov/epr> (04 April 2003).

landfills and Florida might also consider doing the same (EPA 2000, 14). South Carolina is considering a fee on the purchase of new electronic equipment containing CRTs (such as televisions and computer monitors) to help develop a state infrastructure for the recovery and recycling of scrap electronic equipment. In Vermont, manufacturers of certain products containing mercury are told to label these products for sale in the state (EPA 2000, 15). Some states are looking at ways to engage producers of electronic products in the collection and recycling of these products at their end-of-life. New York has proposed take-back legislation for electronic equipment that would require manufacturers to establish collection and/or disassembly centres for recovery of at least 90 percent of the waste equipment. Minnesota initially proposed a stewardship policy that would mandate producer responsibility for CRTs and some other products. The state, however, is currently investigating the degree to which voluntary assistance partnerships with industry can address this waste stream. And New Jersey is establishing an electronics recovery program in Union County to demonstrate how a residential collection program might be economically viable.

3. COMPARATIVE EVALUATION

In the face of differing political cultures, different opinion-forming processes, and varying legislative and political procedures, a detailed, comparative evaluation of the legislative and political approaches in Germany, the European Union, Japan, and the United States is neither achievable nor appropriate. But general tendencies can be drawn, taking into account the economic and social significance of IT in these countries, and especially the PC sector, considering the present unsustainable production, usage, and disposal of computers and their peripherals.

With one exception—the United States—all countries under analysis here have identified the waste problem and initiated countermeasures. The U.S. government has only put its focus on predominately harmful and polluting wastes, whereas clear waste reduction attempts are not on their agenda—a continuation of the U.S. federal government's *laissez faire* mentality. Only recently, this policy was illustrated by the attitude of President George W. Bush regarding the Kyoto Accord, the international climate agreement, which the U.S. failed to support. Apparently, this attitude is also applicable to numerous other areas, such as the waste of raw materials through unsustainable mass production and mass consumption practices.

Also remarkable is the fact that computer monitors from private households are not classified as harmful waste (unless a household produces the unlikely quantity of 100 kilograms worth of monitors a month). But in the face of increasing numbers of PCs in private households and a simultaneous decline of useful PC life, the growing quantity of obsolete monitors from private households alone will demand re-classification. In the United States the lifespan of a new PC is (in contrast to Germany) not primarily driven through depreciation, but through financing and leasing agreements. Most of these agreements are for 36 to 60 months, and depreciation (for tax purposes) is determined based on these time periods. For companies this policy does three things: (1) it extends their credit and ability to borrow, so their cash reserves (cash flow) are not depleted when major equipment upgrades are required; (2) most financing and lease agreements are structured with buy-back incentives. This allows manufacturers to maintain a steady clientele; and (3) companies are not burdened with trying to figure out how to dispose of obsolete equipment. Consequently, and in contrast to Germany, the taxation system in the United States is not a major obstacle towards the re-use and refurbishment of PCs.

With regard to the sustainable development paradigm, national governments must take appropriate measures that take future developments into consideration. Otherwise, especially since the mountains of IT waste continue to grow, they will further fuel the hypothesis of many scientists that governments are only being reactive. Additionally, such policies feed the perception of a lack of ability to steer and control, and thus to form the necessary steps to move towards sustainable development. Nevertheless, it will be interesting to see in the following sections whether certain other approaches (e.g., industry and NGOs) will give reason for such a predominately reactive attitude. Possibly, new framework conditions—through globalisation, for instance—will lead to dynamic innovation in industry and NGOs, so that the mandate of governments is increasingly more to only channel developments and provide the necessary framework conditions.

In contrast to the United States, Japan, which is poor in natural resources and confronted with space limitations, has initiated various approaches to reduce the mountains of waste being generated, including waste from the IT sector and, especially, the PC sector. One key objective is to increase the efficiency of resource use. This will not only help to solve the problem of waste, but also to reduce the dependency on imports of natural resources, thus reducing costs. Consequently, Japan's attempt at realising a recycling-oriented society is by focusing on sustainable development, although concrete measures to channel the enormous material flows from the IT sector (and the PC sector in particular) have not yet been tabled. Furthermore, the simple promotion of recycling involves the risk of low quality, mass recycling. Although the re-use and repair of equipment is also considered important, the overall policy is recycling-oriented. Without specific regulations and a clear, practical definition of "recycling," the legislative requirements could be simply met through re-using waste. Thus, for example, incinerating plastic waste (considered an excellent fuel) could officially be called recycling. Nevertheless, the policies of the Japanese government give the impression that they have taken a future-oriented, holistic problem-solving approach, although the framework setting does not necessarily lead to practical implementation.

The German national laws, in direct connection with the new directives of the European Union, form the most comprehensive and concrete laws (regarding WEEE and PCs), in comparison with Japan and the United States.

But the ambitiousness of the quotas set also gives reason for industry to fear. The WEEE Directive will give final owners of equipment at its end-of-life the right to return goods to distributors free-of-charge. Suppliers will be responsible for providing the infrastructure to collect such equipment and for certain phases in the waste management of their products. For businesses operating in the European Union, the impact is clear—measures must be taken to comply with the new regulations. But the rather large, heterogeneous market within the E.U. will require enormous investments to achieve efficiently functioning take-back and recycling systems, so the regulations may become a large financial burden—even endangering the existence of small and medium-sized companies that dominate the industrial structure of many E.U. countries, such as Germany, Austria, Denmark, etc. On the other hand, national systems to deal with WEEE, where costs are allocated to individual manufacturers based on the costs arising from the return of their own products, support the development of innovative technologies in the design, production, and dealing with the end-of-life phase.

4. OTHER INITIATIVES THAT ENCOURAGE A SUSTAINABLE PC LIFE CYCLE

Apart from governmental and legislative initiatives, approaches by industry and other sectors (examined in the following section) also contribute to the development of policies that encourage a more sustainable PC industry.

4.1 Germany and Europe

4.1.1 Working Groups

The working group on cathode ray tube recycling was formed in Germany upon the initiative of the ZVEI's trade association for "leisure" electronics, which arose against the background of legal and political moves towards the recycling of electronic appliances. Experts representing the manufacturers of cathode ray tubes and cathode ray tube glass in the European Union (including Matushita, Philips, Samsung, Schott, Sony, and Thomson) are members of this group. The working group set itself the goal of examining the requirements and starting principals for using recycled cathode ray tube glass, particularly glass containing lead, in the production of glass for a cone, and to develop a recycling strategy which can be implemented in practical terms (ZVEI 2000). The standardisation of cathode ray tube glass is also relevant for the producers of monitors for PCs and laptop displays. Producers in these areas should commit themselves to a similar initiative (Behrendt 1998, 97).

The CYCLE working group, mentioned earlier, was formed within the ZVEI of the trade association for information technology (VDMA)[24] against the background of the

[24] VDMA stands for *Verband Deutscher Maschinen- und Anlagenbau,* the Association of German Mechanical and Systems Engineering.

KrW, which came into force in 1996. The members of the CYCLE group are producers and commercial enterprises who hold around 80 percent of the IT market. The goal of this group is to avoid a legal directive on the take-back and recycling of IT products by reaching a voluntary agreement first. Until now, the concept of a voluntary agreement was not successful due to the question of distribution of costs. The local authority districts and regions claim that the industry should also assume the cost of the collection, transportation, and initial sorting of the waste appliances. The range of products involved constitutes a further weakness of the proposed agreement, as it only represents 7 percent of the total amount of electronic waste generated (Ebeling 1996; Behrendt 1998, 98f).

The project Care Vision 2000 was initiated by numerous European businesses in the IT sector, led by Sony Europe. It targets the production of electronic appliances, which are long-lasting, and suitable for refurbishing and for recycling. A key aspect of this is the creation of an information system intended to accompany the products and to make product-specific data available to the required parties (Behrendt 1998, 106).

Within the framework of the Information Society Technologies Programme of the European Commission, the project Digital Europe: E-business and Sustainable Development is being carried out by the Forum of the Future (U.K.), the Fondazione Eni Enrico Mattei Institute (Italy), and the Wuppertal Institute for Climate, Environment and Energy (Germany). The aim of the project is to quantify to what extent e-business can contribute to dematerialisation and resource efficiency, what impacts e-business has on the social responsibility of enterprises, and what structural and regional changes can be expected through e-business (e-commerce and e-work). In cooperation with companies, users, and government organisations, the impacts at the micro-level will be analysed and classified according to the e-business platforms of "business-to-business," "business-to-consumer," and "business-to-government" through eight case studies; and trends for the macro-level will be determined with the help of scenario techniques. The project aims to work out optimisation strategies for decision-makers in politics and enterprises on the subject of a sustainable e-Europe and e-business.[25]

4.1.2 Other Strategies

Leasing has long been discussed as a business and political method of reducing environmental damage. The core idea is that consumers purchase certain services that products provide through a lease, instead of buying the actual product; thus the product remains the property of the producer. Leasing enterprises have a vested business interest in the extension of the period of use of a product and using it for its full technical lifespan. The concept of leasing could become problematic, though, if it causes an additional demand for certain products.

[25] Information supplied by Justus von Geibler (Wuppertal Institute, April 2001) and Michael Angrick (German Environment Agency, April 2001).

Various leasing companies have joined together to form the Bundesverband Deutscher Leasinggesellschaften (German Association of Leasing Companies). Some specialise in supplying offices and offer leasing plans for information technology products. A number of other producers (Rank Xerox, Hewlett-Packard, IBM) have formed subsidiary companies that offer leasing, or they offer it themselves. The main products leased out are mainframe computers, workstations, scanners, printers, large copiers, and telephones. In Germany in 1996, the turnover from leasing in the area of IT was around U.S.$3.4 billion (E.U.€3 billion), which amounts to 11.4 percent of the total leasing volume (Behrendt 1998, 187).

Moreover, product-sharing is the start of an innovation to encourage more intensive use of products. The most well-known example is that of car-sharing. In Switzerland, there have been commercial attempts to achieve a form of product-sharing in the area of consumer goods for some years. Leasing programmes for personal computers could not be realised due to lack of sufficient demand. In Germany, the official registered association, Öko-Stadt-Rhein-Neckar e.V. (Eco-City Rhein-Neckar), maintains a shared-use network of consumer goods.

Additionally, the BUND (Friends of the Earth) in Germany provides information about "green" computers. On a special list, users can check on which products are less polluting. The checklist refers to criteria like eco-labeling, power consumption, electromagnetic radiation, pollutants, recycling, and packaging.

Furthermore, a fair number of IT-related eco-labels exist in Germany and the European Union (described by Saied and Velasquez in Chapter 8), like the TÜV ECO Circle 2000 (2000), the European Eco Label for PCs and notebooks (2000), or the German Blue Angel (2000) mark for several IT products. The European Eco Label for PCs and Notebooks and the German eco-label TÜV ECO Circle 2000 require a supplier's declaration, according to the European Computer Manufacturer's Association (ECMA) Technical Report 70 (TR 70).[26] It is difficult to measure the market effects of eco-labeling, but obviously they are relatively well known with the German population.

4.1.3 Regional and Private Initiatives

There are many regional initiatives that upgrade used computers received from organisations for use in schools. These projects are often connected with employing jobless people and are funded by Employment Programs (ABM) and other public institutions in Germany. The following are some examples:

- ECO-PC Berlin, established in 1997 by the non-profit organisation KirchBauhof, upgrades used computers received from enterprises for use in schools at cost price. The services provided to the companies include transport, deletion of data, and the recycling of the computers, which, for them, are unusable. Schools receive the computers at cost price after they have been checked and upgraded for multimedia capability.

[26] European Association for Standardising Information and Communication Systems (see Dworak 2000)

- MookWAT-PC and Nutzmüll e.V. are similar kinds of projects in Hamburg.[27]

4.2 Japan

In December 1999, the Japanese Electronic Industry Development Association (JEIDA) framed a voluntary action programme for reducing, re-using, and recycling computers. This plan produced operational targets to promote, on a voluntary basis, the transition towards a resource recycling, or "zero emissions," society. According to this plan (and thus in order to achieve the "3Rs" for computers), self-governing bodies, retailers, and consumers are required to accept responsibility for making maximum use of the existing schemes of companies through collection systems. Up until March 2000, 24 IT producers had agreed to support this plan and approved the common goal of a 60-percent resource re-use rate by 2005. This rate is based on a calculation dividing the number of units re-used (e.g., used goods, parts, and materials) by the number of units of re-used and recycled goods.

Based on their Environment Charter, Fujitsu, one of Japan's biggest PC producers, set a goal of 90 percent recycling by 2000. As the following figure indicates, this goal was already achieved in 1999.

Nevertheless, only 15 percent of the total products that Fujitsu produces are collected, most of which comes directly from inside the company, will only a small fraction from general households. Additionally, Fujitsu started a refurbishing service for enterprises and introduced a re-use service; but copyrights and warranties on the quality of goods are certainly obstacles to the promotion of re-use

Enforced through the Environmental Impact Assessment (EIA) programme, which was started in the 1970s, IBM Japan has made considerable effort to have less impact on the environment. In 1991 they initiated the Environmentally Conscious Products (ECP) programme in which they promoted design for the environment (DfE), recycling technologies, and standards of measures for environmental awareness. For 2000, the company set the following goals, especially for its new notebook computers:

1. Double the recycling rate from 1995 standards
2. Decrease the time for dismantling by 40 percent
3. Realize a 15 percent decrease in the amount of packing materials
4. Improve on energy-saving features and levels from previous models

4.3 United States

To boost donations and the re-use of computer equipment for schools, the U.S. Congress expanded tax incentives for private companies that donate computer technology, equipment, or software to schools by passing the 21st Century Classroom Act for Private Technology Investment, a provision of the Taxpayer Relief Act of 1997

[27] <http://www.nutzmuell-hh.de> (04 April 2003) in German; <http://www.mookwat-pc.de> (04 April 2003) in German.

(EPA 2000, 5). Large companies that donate computers and related equipment to public or private schools are able to deduct their full purchase costs from their adjusted gross incomes. A clause in the provision prevents the dumping of outdated equipment (e.g., electronic products must be no more than two years old at the time of donation).

Additionally, there are several U.S. Web sites which function as platforms for donations, leasing, recycling, re-use, and product-sharing.[28]

4.3.1 Industry

Extended product responsibility practices are constantly evolving as industry works to achieve resource use efficiency and pollution prevention. A sampling of businesses implementing EPR is provided below.[29]

Sony Electronics announced in October 2000 that it is teaming up with the Minnesota Office of Environmental Assistance (MOEA) and Waste Management Inc. (WMI) to establish a take-back and recycling program for Sony's electronic products. Under the program, consumers throughout Minnesota can return all Sony electronic products free-of-charge. Under a five-year agreement with WMI and MOEA, Sony will subsidize the take-back and recycling of its products in Minnesota until the process becomes cost-effective. Sony hopes to expand the program nationwide within five years.

Dell Computer Corp. is manufacturing a line of professional-level computers that are completely recyclable as part of a move towards an entirely "green" product range. Dell's OptiPlex PCs meet the stringent standards for Germany's Blue Angel eco-label.

Monsanto's production facility in Louisiana has been leasing computer equipment from Dell Computer Corp. in an arrangement that not only reduces waste for Monsanto but also consistently provides them with high-quality computer workstations (EPA 1998, 8).

Gateway Country stores (more than 260 locations in the United States) give discounts on new PCs to individuals who donate functioning 386-class or better computers of any brand to a non-profit organization called Goodwill.[30]

In 1999 alone, the computer recovery operations at the Public Service Enterprise Group's (PSEG) Resource Recovery Centre in New Jersey prevented nearly 54,500 kilograms of electronics, or the equivalent of more than 1,500 desktop computer systems, from entering the waste stream. Of this amount, PSEG donated $220,000 in equipment to more than 80 organizations and sold $105,000 in equipment to more than 300 customers, avoiding almost $57,000 in disposal costs.

[28] EPA (2000) provides an overview of such Web sites.

[29] More initiatives and regional efforts are provided by EPA Online <http://www.epa.gov/epr>.

[30] <http://www.goodwill.org>

5. CONCLUSION

Although IT industries still account for a relatively small share of economic output, the contribution to real economic growth is relatively high. Impressive growth rates reflect the growing significance of the IT sector for national economies. Furthermore, by the end of this decade a majority of employees will be employed in the information sector (EITO 2002). Following the growing production and sales of IT products, the environmental load of PCs is also growing considerably. High energy use, possible health effects due to exposure to particular substances used in computer production, the hazardous substances used in PCs themselves, the ineffective and inefficient consumption of non-renewable natural resources, and finally, the growing mountains of PC-related wastes are the five main issues on the current environmental agenda regarding the life cycle of PCs. Many of the problems associated with these issues remain unsolved. In some cases, as they have not even been identified by some key actors yet, such as legislators, they have thus not made it onto the political agenda.

The obvious trend towards even smaller, harder-to-recycle IT products, as described by Matthews and Matthews in Chapter 1, will probably lead to an intensification of the environmental problems. The increasing stream of obsolete products officially estimated is leading to a flourishing of reclamation business for IT products (ibid.), which will become strategically valuable as scale efficiencies are gained and more public perception of the problem is achieved. Consequently, these new businesses are certainly welcomed and, furthermore, politically intended once mandatory take-back goes into effect. But under the paradigm of sustainability this should only be transition phase. Hieronymi and Schneider argue in Chapter 4 that product return costs should not be allocated to a product or to a product group, but should be imposed on individual manufacturers based on the actual costs arising from the return of their products. This will lead to cost-efficient, high-quality recycling through competition between product-return systems, and thereby also the need to better design products to improve re-usability, recyclability, and the respective innovative technologies, as shown by Klatt in Chapter 11.

The current set of policies in the European Union/Germany, Japan, and United States are predominately concerned with mitigation of the IT waste problem through a mandatory take-back of obsolete products targeting increased recycling quotas. But these can lead to an ever-expanding recycling system that pays no attention to reclamation and reprocessing possibilities, with everything finally ending up in the same disposal process and inhibiting any meaningful recycling. The rather inefficient recycling of all packing materials predominately implemented through the monolithic consortium with the Green Dot (Grüner Punkt) of the Duales System Deutschland GmbH is an example of the danger. The main target of the Green Dot, namely to reduce the enormous mountains of waste, has not yet been achieved. The German Association of Plastics Recycling has come to the conclusion that the necessary separation of the various kinds of plastics is impossible under this system, resulting in low-quality plastic products and the replacement of fossil fuels through incineration, for example, but the Duals System Deutschland GmbH defines all this as recycling (*Geo Magazin* 2000).

In order to learn from the problems with these systems and pave paths in the opposite direction—and thus to effectively implement the polluter-pays principal by rewarding those companies continuously improving the environmental performance of their products—policymakers have to further develop incentives that impose the costs for take-back on the individual manufacturer. One key approach towards this is reducing subsidies on waste disposal through dumps, incineration plants, etc., through long-term licensing and ensuring an effective use of capacities financed substantially with public funds. In addition, a critical but comprehensive public assessment of the existing recycling systems, such as for packaging, needs to be done in order disclose the failures of the past and to debunk alleged successes. In Germany, for example, experts from the sciences and administration agreed some time ago, behind closed doors, that Germany's present waste disposal system is neither effective from an economic point-of-view, nor realizing the targeted environmental benefits. But after more than a decade of extensively promoting and financially supporting these approaches, nobody is willing to take political responsibility and explain to the public: "We have been wrong!" On the contrary, there is a continuation of the rather reactive attitude, waiting for industry to take responsibility, thus also shifting the role to actors outside of government. This feeds theories of a growing inability to govern.

The lack of a long-term strategy for sustainability and questionable policy formulation processes also becomes obvious, shifting the focus away from the end-of-life phase to the production of PCs. Thus Williams highlighted in Chapter 3, for example, the uncertainty regarding the health effects for semiconductor workers due to the lack of medical studies Although the environmental impacts of lead and mercury have been considerably analysed for various product groups such as tubes and batteries, this knowledge has yet to be applied to assess the risk associated with heavy metals in electronics and especially IT. But thus, a substantive basis to justify the decision of the European Union to ban these substances does not exist, and there is no clear basis to continue the formulation of future measures to deal with the environmental issues associated with electronics. In this regard the European Union's policy formulation is made on assumptions, leaving consumers, producers, and employees with much uncertainty, especially considering that some companies are labeling their products as eco-friendly, as explained in Chapter 8 by Saied and Velasquez.

As for the use phase of PCs, Cole explained in Chapter 7 that a lack of awareness on the part of the user is an obvious key obstacle in realizing PC energy savings, which suggests that the education of users and systems managers is critical. Firms and government agencies need to respond by ensuring that the formulation and implementation of policy clearly identifies and puts to rest the myths and misconceptions which prevent the user from taking steps to reduce PC energy consumption. But again, as already described for the end-of-life and production phase, governmental policy approaches towards this aim are not yet obvious or presently on political agendas.

It is apparent that the current set of policies in Germany/European Union, Japan, and the United States are most concerned with recycling and do not really do much to address the extension of PC life, reducing harmful effects during production, energy consumption during usage, or the reduction of material inputs per se. And there are even political decisions with rather absurd results from the perspective of sustainability. One example is the taxation rules on new PCs that lead to a lifetime of at most three years and hinder, in practice, the utilisation of refurbished PCs. The tax systems should assess refurbished PCs at least equal to new ones. One could even imagine, for example, that through developing an incentive system for those trying to prolong the lifespan of their IT products the consequence would be depreciation according to the age of the machines. Such a system would easily create additional impetus for manufacturers to reconsider design approaches and make products more easily upgradeable.

In addition, various strategic approaches towards sustainable development have another key approach in common—dematerialisation, through developing a service-oriented society, based on leasing or rental of the services of provided by PCs and other information and communication technology products (Robèrt 2002). Consequently, leasing and renting could also lead to certain credits for selected products; but it has to be investigated further. Simultaneously, national and international legislation should favour the smooth transfer of software rights to secondary owners, to bring them into the position of selling PCs that might not have the latest developments, but easily fulfill the original purposes of the client, such as word-processing and e-mail.

In summary, one must confess that the present policies are not sufficient enough to manage the environmental impact of PCs in a sustainable way, as is the understanding of some of the interdependencies and dimensions of the PC life cycle. Granted, a few promising developments have occurred, such as the WEEE and ROHS directives in the European Union, but they also will require further political incentives to accomplish any real innovation.

The case of the PC can also be seen as another example of a stalemate in the development of practical approaches for sustainable development. Neither Germany/the European Union, the United States, nor Japan have yet developed an environmental policy of the so-called second generation, taking the three dimensions of sustainable development—ecological, economic, and social—mutually into account. Policies are still dominated by curative countermeasures that uncouple development and social satisfaction, instead of a holistic, preventive strategy towards the dematerialisation of products and services. The often-expressed political will towards this aim must be backed-up by efficient and constructive cooperation with all other major actors' groups, which is a new form of an old challenge to politics in a globalized world. It will require a reorganization of political decision-making processes, but also offers opportunities for more influence on a transnational level.

REFERENCES

Behrendt, S. 1998. Innovationen zur Nachhaltigkeit. Ökologische Aspekte der Informations- und Kommunikationstechniken. Frankfurt/Main: Springer.

Bundesanstalt für Arbeitsschutz und Arbeitsmedizin (BAA). 1997. Erhebung der Belastungssituation beim Recycling von Elektronikschrott. Dortmund.

Bundesministerium für Umwelt (BMU). 2001. Gesetzgebung Altgeräte. <http://www.bmu.de/sachthemen/abfallwirtschaft/bmu-stadt/elektro/detail/alt_gesetz.htm> (23 October 2002).

Bundesratsdrucksache 638/98 (in German).

Commission of the European Communities. 2000a. Proposal for a directive of the European Parliament and the Council on restriction of the use of certain hazardous substances in electrical and electronic equipment. 2000/0159(COD), COM 2000 347 provisional, Brussels.

———. 2000b. Proposal for a directive of the European Parliament and the Council on waste electrical and electronic equipment. 2000/0158(COD), COM 2000 347 provisional, Brussels.

Donicht, Ch. and U. Selent. 1996. Das neue Abfallgesetz. Institut für ökologische Wirtschaftsforschung (IÖW) Berlin.

Dworak, Th. and M. Kuhndt. 2000. Return of used IT products, reuse, recycling. Policy paper of the Wuppertal Institute for Climate, Environment and Energy.

Ebeling, A. 1996. Abgeschreddert. Das Elend mit dem Computerschrott. In *c't* (4):124–129.

Environmental Health Centre (EHC). 1998. Electronic Product Recovery and Recycling Conference. Summary Report.

Environmental Protection Agency (EPA). 1998. WasteWise update: Extended product responsibility. <http://www. epa.gov./wastewise> (20 March 2003).

———. 2000. WasteWise update: Electronics reuse and recycling. <http://www.epa.gov./wastewise> (20 March.2003).

European Information Technology Observatory (EITO). 2002. European Information Technology Observatory 2002. Frankfurt/Main.

Europäische Kommission. 1996. Gemeinschaftsrecht im Bereich des Umweltschutzes – Abfall (Vol. 6). Brussels.

Geo Magazin. 2000. Grüner Punkt für grünes Gewissen. *Geo Magazin*, Juli 2000. <http://www.geo.de/GEO/wissenschaft_natur/oekologie/2000_06_GEO_07_gruener_punkt/?SDSID=-> (10 April 2003).

Grote, A. 1999. Kreisläufer. Die Elektronikschrott -Verordnung kommt nicht Voran. In *c't* (8): 90–94.

Japan Times Online. 2003. Post offices to aid in recycling PCs. *Japan Times Online*, 8 April. <http://www .japantimes.com/cgi-bin/getartcile.pl5?nb20030408a9.htm> (10 April 2003).

Konrad Zuse Internet Archive. 2001. <http://www.zib.de/zuse/English_Version/index.html> (01 April 2003).

Kuehr, R. 1996a. Tokyos Müllmanagement in Zeiten zunehmender Raumnot. In *Japan 1995/96*, ed. Manfred Pohl, 56–71. Hamburg.

———. 1996b. Dreamland of waste. Tokyo's waste-management in times of land shortage. In *Global environment change: Human and policy dimensions* (Oxford) 6(2): 173–175.

Kunig, Ph. et al. 2003. Kreislaufwirtschafts- und Abfallgesetz. München: Beck.

Nikkei Net Interactive. 2003. Business glossary: Environment. <http://www.nni.nikkei.co.jp/FR/TNKS/TNKSHM/glossary/env_02.html> (10 April 2003).

Patton-Hulce, V. 1995. *Environment and the law. A dictionary*. Santa Barbara: ABC-CLIO.

Robèrt, K.-H. et al. 2002. Strategic sustainable development. Selection, design and synergies of applied tools. *Journal of Cleaner Production* 10(3): 197–214.

Sozialdemokratische Partei Deutschlands (SPD). 2003. Koalitionsvertrag—Erneuerung, Gerechtigkeit, Nachhaltigkeit. <http://www.spd.de/servlet/PB/show/1023294/Koalitionsvertrag.pdf> (01 April 2003).

Thomsen, S. 1997. Produktverantwortung: Rechtliche Möglichkeiten und Grenzen einer Kreislaufwirtschaft. Nomos. Baden-Baden.

Umweltbundesamt (UBA). 1996. Stand der Entsorgung von elektrischen und elektronischen Kleingeräten in der Bundesrepublik Deutschland. Nr. 96/61. Berlin.

Welker, A. et al. 1996. Waste from electrical and electronic equipment: Producer responsibility: A view of initiatives in the EC. *European Law Review* (12): 341–344.

Wuppertal Institute. 2000. <http://www.wupperinst.org/energie/werkstatt/i-net.htm> (07 December 2001).

ZVEI. 2000. Entsorgung von Elektro- und Elektronikaltgeräten. Fakten und Argumente.

APPENDIX 1

Further reading on policies in Japan

1. Denki/Denshi Kiki no Recycle Gijutsu (Toushiba Kankyou/Recycle gijutsu kenkyuuka), Tokyo 1999.
2. Energy to Kankyo – shigen, kankyo, recycle to chikyuondanka – (Sangyougijutsu Kaigi).
3. Shigen, Kankyo, Recycle – junkangata sangyou keizai system no kakuritsu – (Sangyougijutsu Kaigi).
4. Haikibutsu to Recycle no Koukyouseisaku (Chuo Keizaisya Apr 2000).
5. Sangyou Recycle Jiten – Shigen junkangata syakai & zero emission wo mezashite – (Sangyou Recycle Jiten Hensyu Iinkai Jan 2000).
6. Personal computer white paper 1999–2000 (Computer Age Sha Aug 1999).
7. Shiyouzumi Computer no kaishu shori recycle no joukyou ni kansuru tyousa houkokusyo (JEIDA Mar 2000).
8. Nikkei Ecology (Monthly magazine 2000).

CONTRIBUTORS

ANNE BRINKLEY (1955) – IBM USA – Corporate Environmental Affairs; program manager, Design for Environment and Life Cycle Assessment; staff member, IBM Engineering Center for Environmentally Conscious Products (since 1991); active in ISO Technical Advisory Group 207.
Contact: annebb@us.ibm.com

DANIELLE COLE (1976) – University of New South Wales, Australia – Director, Green Office Program; co-ordinator, *unswitch* (UNSW energy conservation program); and initiator of the Australian Campuses Towards Sustainability (ACTS) Network.
Contact: d.cole@unsw.edu.au

KLAUS HIERONYMI (1952) – Hewlett-Packard, Europe – General manager, Environmental Businesses Europe (HP). Responsible for Environmental Management Europe (since 1999), including monitoring environmental legislation and market trends, as well as HP's European take-back and recycling operations; held various positions in sales and marketing at HP Germany, Europe and Asia; studied natural sciences and business management in Germany and Austria.
Contact: Klaus.Hieronymi@hp.com

REINHARD HÖHN (1952) – IBM Germany – Manager, Environmental Affairs and Product Safety, IBM EMEA Central Region; chair, environmental council, BITKOM (German association of information and communication technology); chair, Energy Task Force, EICTA (European Information Systems Communication Technologies and Consumer Electronics Industry Association); and chair, Technical Committee 38 on product environmental aspects of ECMA-International.
Contact: hoehn@de.ibm.com

STEFAN KLATT (1964) – MAN Nutzfahrzeuge AG, Germany – Engineer; department manager (since 2001), Institutional Relations and Transport Policy; project leader, Duales System Deutschland AG (the Green Dot); leader, end-of-life vehicles industry project (German Car Recycling Association); and consultant at bvse-Bundesverband Sekundärrohstoffe und Entsorgung e.V., the German recycling association for small- and medium-sized businesses.
Contact: Stefan_Klatt@mn.man.de

RUEDIGER KUEHR (1970) – United Nations University, Zero Emissions Forum, Germany – Political and social scientist; European Focal Point of United Nations University's Zero Emissions Forum (since 1999) <www.unu.edu/zef>; international co-ordinator and secretary, Alliance for Global Eco-Structuring (AGES) under UNEP's Cleaner Production Network; research specialist, The Natural Step (TNS) Stockholm; and freelance policy-consultant to various international organisations, national governments, and companies.

Contact: kuehr@online.de

DEANNA H. MATTHEWS (1972) – Carnegie Mellon University, USA – Research associate, Green Design Institute and the Department of Civil and Environmental Engineering. Research includes the engineering and economic feasibility of materials recycling, corporate environmental performance and metrics, and environmental management systems; has worked in the electronics industry and environmental consulting.
Contact: dhm@cmu.edu

H. SCOTT MATTHEWS (1970) – Carnegie Mellon University, USA – Assistant professor, Civil and Environmental Engineering, and Engineering and Public Policy; research director, Green Design Institute, Carnegie Mellon University, Pittsburgh, PA, USA. Research interests: life cycle assessment of products and services, socio-economic and environmental implications of information and communications technology, and energy and environmental implications of infrastructure systems.
Contact: hsm@cmu.edu

HARALD PODRATZKY (1955) – Fujitsu Siemens Computers, Germany – Mechanical engineer; responsible for environmental management; has worked a number of years in planning and optimizing production processes; introduced environmental management to the development and production of PCs (1995); also works as an auditor of quality and environmental management systems.
Contact: Harald.Podratzky@fujitsu-siemens.com

MOHAMED SAIED (1964) – United Nations University, Global Environment Information Centre, Japan – Research associate, Global Environment Information Center; Ph.D., Graduate School of Information Systems, University of Electro-Communications, Tokyo; taught at the University of Tunis II, Tunisia; former research associate, Tokyo Medical and Dental University. Research interests: human/computer interaction, network security, electronic education, computer applications for the handicapped, and information technology for sustainable development.
Contact: saied@hq.unu.edu

JOSEPH SARKIS (1963) – Clark University, Graduate School of Management, USA – Professor, Operations and Environmental Management; Ph.D., University of Buffalo; has published over 150 articles on topics such as environmentally conscious business practices, technology management, supply chain management, performance management, and multiple criteria decision-making.
Contact: jsarkis@clarku.edu

YUKIHIRO SASAKI (1977) – Mizuho Financial Group, Japan. While a master's degree student at the department of Industrial Engineering at the Tokyo Institute of Technology, he worked as a research assistant for the Information Technology (IT) and Environmental Issues project, United Nations University, Tokyo <www.it-environment.org>.
Contact: y-sasaki@pop06.odn.ne.jp

AXEL SCHNEIDER (1949) – promotion team Wetzlar, Germany – Managing director; graduate, electronics and communication design; supporter of Hewlett-Packard's General Manager Environmental Businesses Europe program.
Contact: axel.schneider@promotionteam.de

GERMAN T. VELASQUEZ (1966) – United Nations University, Global Environment Information Centre, Japan – Coordinator, UNU Global Environment Information Centre, a joint initiative of the UNU and the Ministry of Environment, Japan. Among his publications, he has written on the UN and the Internet, and co-developed "Pangaea," an Internet-based sustainable development gaming simulation, and "Quake Busters," a multi-media software package for disaster awareness education in Japan.
Contact: jerry@geic.or.jp

ERIC WILLIAMS (1965) – United Nations University, Japan – Project coordinator, Environment and Sustainable Development Programme, United Nations University (UNU) Centre, Tokyo; Ph.D., theoretical physics, State University of New York, Stony Brook. Since 2001, he has been conducting research and project management for the Information Technology (IT) and Environmental Issues project at UNU, which endeavors to understand and develop appropriate social responses to environmental aspects of the IT revolution <www.it-environment.org>.
Contact: williams@hq.unu.edu

INDEX

Eco-Efficiency in Industry and Science

1. J.E.M. Klostermann and A. Tukker (eds.): *Product Innovation and Eco-efficiency.* Twenty-three Industry Efforts to Reach the Factor 4. 1997 ISBN 0-7923-4761-7
2. K. van Dijken, Y. Prince, T. Wolters, M. Frey, G. Mussati, P. Kalff, O. Hansen, S. Kerndrup, B. Søndergård, E. Lopes Rodrigues and S. Meredith (eds.): *Adoption of Environmental Innovations.* The Dynamics of Innovation as Interplay Between Business Competence, Environmental Orientation and Network Involvement. 1999 ISBN 0-7923-5561-X
3. M. Bartolomeo, M. Bennett, J.J. Bouma, P. Heydkamp, P. James, F. de Walle and T. Wolters: *Eco-Management Accounting.* 1999 ISBN 0-7923-5562-8
4. P.P.A.A.H. Kandelaars: *Economic Models of Material-Product Chains for Environmental Policy Analysis.* 1999 ISBN 0-7923-5794-9
5. J. de Beer: *Potential for Industrial Energy-Efficiency Improvement in the Long Term.* 2000 ISBN 0-7923-6282-9
6. K. Green, P. Groenewegen and P.S. Hofman (eds.): *Ahead of the Curve.* Cases of Innovation in Environmental Management. 2001 ISBN 0-7923-6804-5
7. J.B. Guinée (ed.): *Handbook on Life Cycle Assessment.* Operational Guide to the ISO Standards. 2002 ISBN 1-4020-0228-9
8. T.J.N.M. de Bruijn and A. Tukker (eds.): *Partnership and Leadership.* Building Alliances for a Sustainable Future. 2002 ISBN 1-4020-0431-1
9. M. Bennett, J.-J. Bouma and T. Wolters (eds.): *Environmental Management Accounting.* Informal and Institutional Developments. 2002 ISBN 1-4020-0552-0; Pb: ISBN 1-4020-0553-9
10. N. Wrisberg and H.A. Udo de Haas (eds.): *Analytical Tools for Environmental Design and Management in a Systems Perspective.* 2002 ISBN 1-4020-0626-8
11. R. Heijungs and S. Suh: *The Computational Structure of Life Cycle Assessment.* 2002 ISBN 1-4020-0672-1
12. M. Bennett, P.M. Rikhardsson and S. Schaltegger (eds.): *Environmental Management Accounting - Purpose and Progress.* 2003 ISBN 1-4020-1365-5; Pb: ISBN 1-4020-1366-3
13. R.U. Ayres, L.W. Ayres and I. Råde: *The Life Cycle of Copper, Its Co-Products and Byproducts.* 2003 ISBN 1-4020-1552-6

KLUWER ACADEMIC PUBLISHERS – DORDRECHT / BOSTON / LONDON